Synthesis Lectures on Digital Circuits & Systems

Series Editor

Mitchell A. Thornton, Southern Methodist University, Dallas, USA

This series includes titles of interest to students, professionals, and researchers in the area of design and analysis of digital circuits and systems. Each Lecture is self-contained and focuses on the background information required to understand the subject matter and practical case studies that illustrate applications. The format of a Lecture is structured such that each will be devoted to a specific topic in digital circuits and systems rather than a larger overview of several topics such as that found in a comprehensive handbook. The Lectures cover both well-established areas as well as newly developed or emerging material in digital circuits and systems design and analysis.

Khaled Mohammed

The Art of RTL Debugging

Strategies and Techniques

Springer

Khaled Mohammed
Cairo, Egypt

ISSN 1932-3166 ISSN 1932-3174 (electronic)
Synthesis Lectures on Digital Circuits & Systems
ISBN 978-3-032-15879-6 ISBN 978-3-032-15880-2 (eBook)
https://doi.org/10.1007/978-3-032-15880-2

This Springer imprint is published by the registered company Springer Nature Switzerland AG
The registered company address is: Gewerbestrasse 11, 6330 Cham, Switzerland

To everyone who has helped me understand, who gave me strength when I felt weak and stood by me through every challenge, thank you for being my endless source of energy and support. This journey wouldn't have been possible without you.

With admiration and love,

Preface

This book presents a comprehensive overview of RTL (Register Transfer Level) debugging, addressing both foundational concepts and emerging technologies. It begins by examining the key issues faced in RTL design, such as timing violations, limited signal observability, and complex verification environments. It then outlines the basic principles of RTL debugging, including simulation techniques, waveform inspection, and the use of assertions and testbenches. Traditional debugging methods are discussed in depth, highlighting techniques like print-style debugging, coverage analysis, and assertion-based verification, along with their respective advantages and limitations. The book concludes with an exploration of cutting-edge approaches involving Large Language Models (LLMs), showcasing how AI can support RTL debugging by assisting in code comprehension, anomaly detection, and bug localization. Together, these chapters provide a clear and progressive understanding of RTL debugging, making the book valuable for both novice and experienced hardware engineers.

Cairo, Egypt

Khaled Mohammed

Competing Interests The author has no competing interests to declare that are relevant to the content of this manuscript.

Contents

About the Author

Khaled Mohammed received his B.Sc. degree in Electronics and Communications Engineering with distinction and honors in 2003 from Ain Shams University, Cairo, Egypt. He obtained his M.Sc. and Ph.D. degrees in Electronics and Communications Engineering in 2008 and 2012, respectively, followed by an MBA degree in 2016. Throughout his career, Dr. Mohammed has contributed extensively to the design and verification of advanced System-on-Chip (SoC) intellectual property cores, including high-speed interconnects and memory interfaces. He has also led several engineering initiatives in hardware emulation, system modeling, and complex SoC integration. Dr. Mohammed has authored and co-authored numerous books, patents, and over 157 peer-reviewed research papers published in leading international journals and conferences. His research interests span 3D integration, IP modeling, hardware emulation, machine learning, artificial intelligence, and Internet of Things (IoT) applications. He is a Senior Member of the IEEE and actively serves as a reviewer for several prestigious journals, including the IEEE Transactions on VLSI Systems, IEEE Transactions on Circuits and Systems II, IEEE Transactions on Semiconductor Manufacturing, IEEE Microwave and Wireless Components Letters, IEEE Transactions on Microwave Theory and Techniques, and the Elsevier Microelectronics Journal.

RTL-Level Design Hazards "A Pre-Debugging Guide"

1

1.1 Introduction

The register-transfer level (RTL) abstraction plays a vital role in the modern VLSI design flow. At this stage, designers use hardware description languages such as Verilog and VHDL, or high-level synthesis (HLS) languages like C and C++, to define the precise behavior of digital systems. RTL design offers a wide design space and significant flexibility, enabling designers to make detailed decisions that will strongly influence the final quality of the ASIC in terms of power consumption, performance, and area (PPA). It is crucial to optimize RTL designs thoroughly during this phase, as addressing inefficiencies or poor-quality RTL in later synthesis stages is extremely difficult and often impractical [1].

In the world of RTL (Register Transfer Level) design, understanding the potential hazards before debugging begins can save time and effort in the long run. Identifying and addressing design issues at an early stage prevents costly debugging sessions later in the development process. This chapter categorizes the most common RTL design hazards, grouping them into higher-level categories such as State Management, Signal Integrity, Design Constructs, and Synthesis Considerations. These categories encompass the typical design errors that arise at the RTL level and offer guidance for avoiding these issues before they become a source of debugging challenges.

To avoid RTL design hazards, it is crucial to address potential mismatches between simulation and synthesis, which can lead to unexpected behavior when transitioning from simulation to actual hardware implementation. Ensuring that the design is synthesized correctly requires that all elements of the RTL code, such as modules, signals, and states, are correctly defined and connected. Simulation can sometimes provide false positives or negatives if the synthesis tool doesn't handle specific constructions as expected. To avoid this, it is recommended to use synthesizable constructs, avoid simulation-only operators

K. Mohammed, *The Art of RTL Debugging*, Synthesis Lectures on Digital Circuits & Systems, https://doi.org/10.1007/978-3-032-15880-2_1

(like === or !==), and ensure that the design is verified with both simulation and synthesis tools to catch these mismatches early.

Connectivity issues are another significant hazard in RTL design. The proper connection of all signals and modules is essential to prevent logic errors and improve the overall integrity of the design. Ensuring correct signal connections and avoiding situations where signals are left unconnected or poorly routed will help prevent issues like race conditions and contention. By thoroughly reviewing the connectivity of the design and verifying the routing paths, designers can catch potential errors that might otherwise go unnoticed in simulation.

Improving coding style is also key to preventing RTL hazards. Writing clean, well-documented, and structured code helps avoid confusion and errors that may arise during the design process. Adhering to consistent naming conventions, minimizing complex logic, and maintaining readability in the code makes it easier to spot issues like incomplete sensitive lists, improper clocking, or unnecessary latches. Additionally, using tools like linting can assist in enforcing best practices and identifying potential hazards before they cause problems during synthesis or simulation. By focusing on these key areas, designers can greatly reduce the likelihood of encountering RTL issues and ensure a smoother transition from design to implementation. Moreover, writing proper comments is an important step towards ease of debugging [2].

1.2 RTL-Level Design Hazards

A hazard is a condition that can cause unreliable/unexpected operation. Figure 1.1 summarizes hazards with RTL design. RTL-level design hazards represent critical challenges in the digital design process that can lead to unreliable or unexpected system behavior if not properly addressed. These hazards typically stem from a range of synthesis and simulation mismatches, coding practices, and connectivity issues that compromise design correctness and performance. One significant source of such problems is the mismatch between simulation and synthesis results, often caused by race conditions, incorrect use of parallel case statements, or constructs that cannot be synthesized. Additionally, using non-synthesizable operators like triple equals, not-equals, or delayed assignments exacerbates the issue by introducing ambiguity between functional simulation and synthesized hardware. Finite State Machines are another common source of RTL hazards, especially when they include unused or unreachable states or contain dead states, which can complicate both verification and synthesis. Clock and reset management also plays a pivotal role in avoiding synthesis problems; using internal clocks or resets, or implementing multiplexed clocks, can introduce timing unpredictability and complicate design timing closure. Coding style further influences the robustness of RTL designs. Care must be taken to ensure data type consistency across operations, particularly with signed and unsigned data. Furthermore, the use of control flow structures like while, forever, or for loops without proper break mechanisms can result in infinite loops or synthesis errors. An

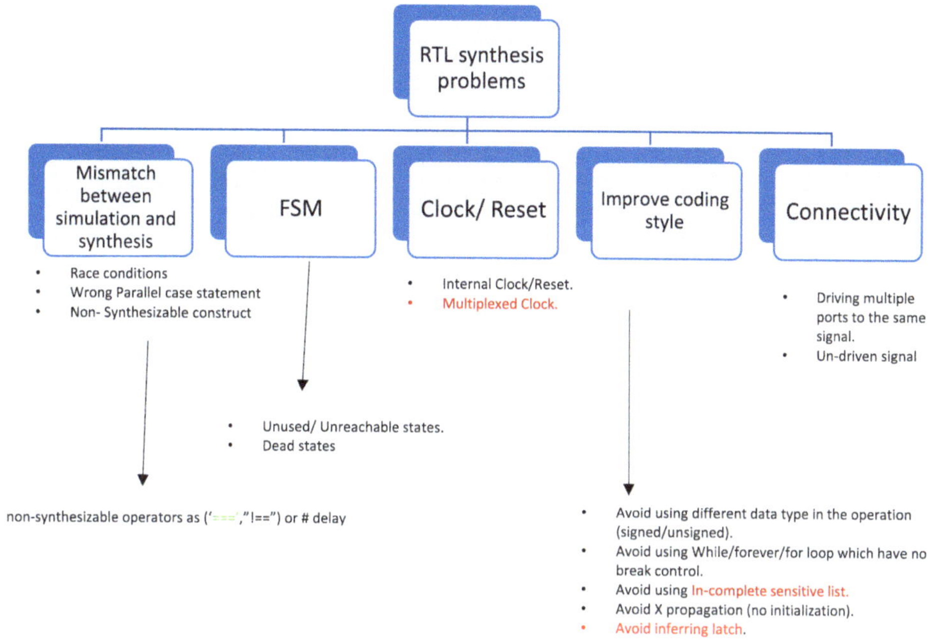

Fig. 1.1 Issue with RTL

incomplete sensitivity list can cause the synthesized circuit to behave differently from its simulation counterpart, while failure to manage unknown states or values (such as those resulting from improper initialization) can lead to undefined behavior due to X propagation. Finally, connectivity issues such as driving multiple ports with the same signal or leaving signals undriven introduce ambiguity in signal control and data flow, leading to malfunctioning hardware. Collectively, these challenges underscore the importance of rigorous RTL design methodologies and thorough validation strategies to ensure reliable and synthesizable hardware descriptions [3–11].

1.2.1 State Management Issues

State management in finite state machines (FSMs) and other control logic is a critical part of RTL design. Many issues arise when the states or transitions are not properly defined, causing problems during simulation or post-silicon operation.

1.2.1.1 Unused and Unreachable States

States in an FSM that are not actively used or reachable can increase complexity and introduce unnecessary design elements that are difficult to debug.

```
always @ (d)
begin
    case (d)
    2'b0x:    begin z = 1'b1; s = 1'b0; end;
    2'bx0:    begin z = 1'b0; s = 1'b0; end;
    default:  begin z = 1'b0; s = 1'b0; end;
    endcase
end
```

Fig. 1.2 Incomplete case statements

1.2.1.2 Dead States

Dead states are unreachable states that never occur during operation but can still cause issues in the design. These can lead to an increased size of the design and can obscure real issues.

1.2.1.3 Wrong Parallel Case Statement

Incorrect or incomplete case statements can lead to incorrect driving of signals, creating bugs that are difficult to detect (Fig. 1.2).

1.2.2 Signal Integrity Issues

Signal integrity refers to the quality and behavior of signals within the design. Problems in this area can lead to unpredictable behavior, timing errors, and difficulty in diagnosing issues.

1.2.2.1 Internal Clock/Reset Issues

Poor synchronization of internal clocks and resets can cause erratic behavior and unexpected failures in the design.

1.2.2.2 Multiplexed Clock

Handling multiple clock domains without proper synchronization can introduce race conditions and timing errors.

1.2.2.3 Multi-Driven Nets

When multiple sources attempt to drive a signal, contention occurs, causing unpredictable behavior in the system.

The variable of data type "wire" can NOT be repeatedly assigned (Fig. 1.3).

The variable of data type "reg" can only be assigned in single always block (Fig. 1.4).

```
assign D = (A)? B: C;
assign D = (B)? A|C: ~A&C;
```

Fig. 1.3 Multiple sources attempt to drive a signal (wire)

Fig. 1.4 Multiple sources attempt to drive a signal (reg)

```
always (*) begin
        if (A)
            D = B&C;
        else
            D = B|C;
end

always (*) begin
        if (A&B)
            D = 1;
        else
            D = 0;
end
```

Multi-driven nets can be created in the RTL by introducing drivers of same or different signal strengths. However, driving a net with multiple signals is not considered as good design practice.

This could lead to a failure in post silicon verification. Many EDA tools do not allow multi-driven nets in the design, and the designers are expected to remove all multi-driven nets from the design.

1.2.2.4 Un-Driven Signals

Signals that are left un-driven may propagate unknown values, leading to simulation or hardware errors. Undriven nets are useless and they generally get deleted by the EDA logic optimization tools.

1.2.3 Design Constructs and Control Flow

Certain RTL design constructs and control flow patterns can create issues that complicate both simulation and synthesis, often leading to incorrect or non-synthesizable designs.

1.2.3.1 Non-Synthesizable Constructs

Constructs like for, while, or forever loops without break control may not translate into valid hardware and can cause failures in synthesis.

1.2.3.2 Avoid Using Different Data Types in Operations

Mixing signed and unsigned data types in the same operation can lead to sign extension or truncation issues, which can create unintended behavior in the design.

1.2.3.3 Avoiding Incomplete Sensitive Lists

Sensitive lists that miss relevant signals can lead to logic that fails to react to changes in critical parts of the design.

1.2.4 Timing and Synchronization Issues

Timing and synchronization errors are pervasive in RTL designs, especially in complex systems like SoCs with multiple clock domains.

1.2.4.1 Race Conditions

When two or more parts of the design attempt to change the value of a shared signal simultaneously, race conditions occur, leading to unpredictable behavior. To avoid, don't write the same signal in different processes, unless you really know what you do.

A race condition is a hazard that occurs in **sequential circuits** when two or more signals arrive at a flip-flop or latch at almost the same time, causing unpredictable behavior. This can result in **metastability** (Fig. 1.5), where the flip-flop's output remains undefined for a brief period (enter an invalid state).

Metastability happens when flip flop inputs change at the same time as the clock. It occurs when violating setup time happens.

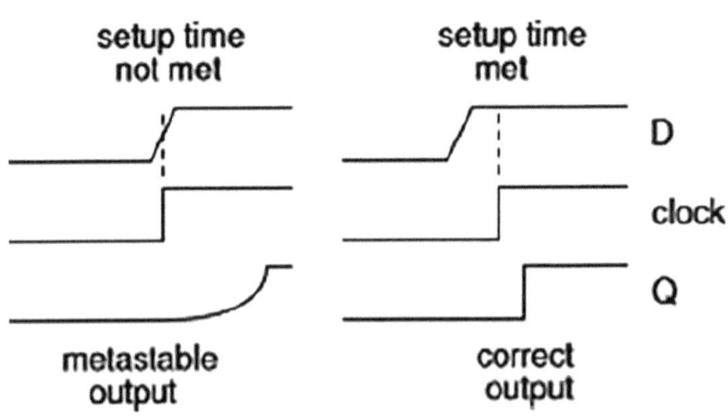

Fig. 1.5 Metastability

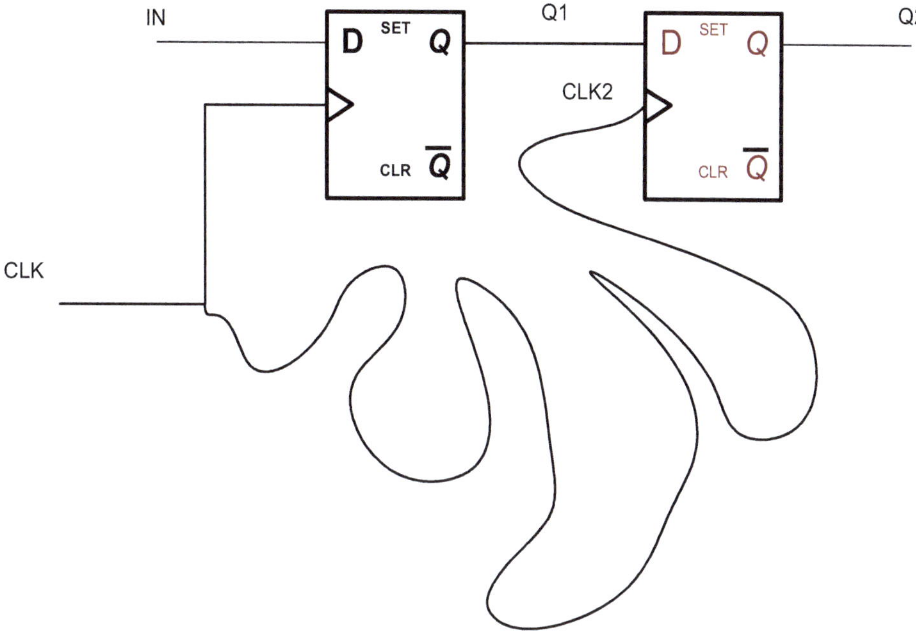

Fig. 1.6 Clock skew

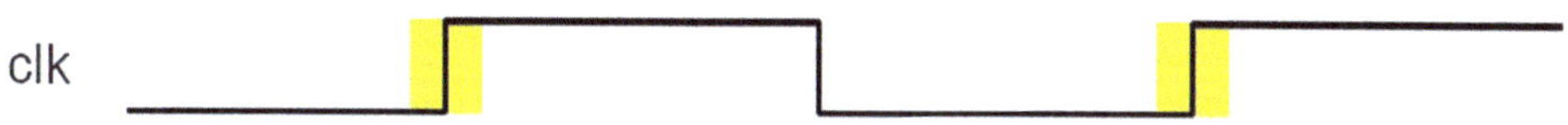

Fig. 1.7 Clock jitter

Assume, few cycles are needed to stabilize data and now one data is under latching process. This data is not yet stabilized within this latch. Now, if you allow one more input to enter what will happen to that data which was under processing. Typically, latch may start processing new input data or may go to an unknown loop state that we call a **metastable** state. in flip-flop-based synchronous designs, **clock skew** or async reset can lead to metastability issues. Asynchronous resets can lead to metastability; using synchronizers and synchronized resets can mitigate this risk.

Clock skew means the clock arrives at different locations at different time. It is measured using eye diagram (Fig. 1.6).

Clock jitter means the clock arrives at the same location at different time. It is measured using eye diagram (Fig. 1.7).

1.2.4.2 Clock/Reset Timing Issues

Poor management of clock and reset signals can cause devices to operate incorrectly, especially during initialization or transitions between different operating states.

```
//Latches inferred because of missing assignments and missing condition
always @ (d)
begin
    case (d)
    2'b00: z = 1'b1;
    2'b01: z = 1'b0;
    2'b10: z = 1'b1; s = 1'b1;
    endcase
end
//Correct2
always @ (d)
begin
    case (d)
    2'b00:   begin z = 1'b1; s = 1'b0; end;
    2'b01:   begin z = 1'b0; s = 1'b0; end;
    2'b10:   begin z = 1'b1; s = 1'b1; end;
    default: begin z = 1'b0; s = 1'b0; end;
    endcase
end
```

Fig. 1.8 An example for interfering latch (non-blocking statement)

1.2.5 Synthesis and Simulation Considerations

Synthesis tools rely on proper RTL code to generate hardware logic, while simulation tools check for correctness during operation. Issues in this category affect both the synthesizability and accuracy of simulation results.

1.2.5.1 Inferring Latches

Latches are often a sign of missing or incomplete logic. They can cause memory elements to retain their value unexpectedly, leading to non-deterministic behavior (Figs. 1.8 and 1.9).

1.2.5.2 Non-Synthesizable Operators

Operators like === or !== are useful in simulation but are not synthesizable, making the design incompatible with hardware implementation tools. Non-Synthesizable Verilog Statements are described in Table 1.1. Some constructs behave differently in simulation and synthesis, which leads to false confidence from simulation-only verification.

1.2.5.3 # Delay Constructs

The use of delay constructs (#) can result in incorrect synthesis or simulation behavior, as these do not correspond to actual hardware delays and should be avoided in RTL designs.

1.2.6 Initialization and Unknown States

Uninitialized or unknown signal states can propagate through a design, leading to difficult-to-trace errors.

Fig. 1.9 An example for interfering latch(blocking statement)

```
//Blocking assignments
always @ (a)
    c <= a or b;
//correct
always @ (*) // (a or b)
    c = a or b;
```

Table 1.1 Constructs not supported in synthesis

#	Constructs not supported in synthesis
1	"Hierarchical name reference not supported" card.resp_gen.device_reg
2	Time: # 580 ns
3	Assign on reg not allowed (but it is ok for wire) reg [15:0] block_cnt = 2
8	"Mixed blocking and non-blocking assignment is not supported." X = 1; X <= 1;
9	Real datatype
10	Initial statement
11	Repeat, while, forever statements
12	Division and modulus operators for variables
13	Non Fixed-size for loops

1.2.6.1 X Propagation

An X value in a signal represents an unknown or undefined state. Without proper initialization, X values can propagate through the design, making it harder to determine the cause of a failure.

1.2.7 Interface and Connectivity Issues

Incorrect interconnects between modules or mismatched interfaces can result in silent failures or undefined behaviors. These are often hard to debug due to their subtle symptoms.

1.2.7.1 Mismatched Port Widths

Connecting signals of different widths without proper truncation or extension can result in data corruption or functional errors.

1.2.7.2 Missing or Incorrect Handshake Logic

For interfaces like AXI, APB, or custom protocols, missing valid/ready or req./ack can cause data loss or deadlock.

1.2.7.3 Incorrect Clock Domain Crossing (CDC)

Failure to properly synchronize signals across different clock domains can lead to metastability, glitches, or random data corruption. Figure 1.10 shows an illustrative example of CDC.

1.2.8 RTL Coding Style and Maintainability

Even if functionally correct, poor coding style leads to lower readability, higher maintenance effort, and more bugs in the long term.

1.2.8.1 Deeply Nested Logic or Conditions

Excessive nesting makes RTL hard to trace and simulate/debug. Flattening logic improves readability and synthesis.

1.2.8.2 Overuse of Macros

While macros are useful, overuse or poor naming can make debugging much harder. It's better to use functions or parameters.

1.2.8.3 Lack of Modularity

Monolithic RTL designs with little reuse or clear module boundaries are hard to integrate or debug in SoC environments.

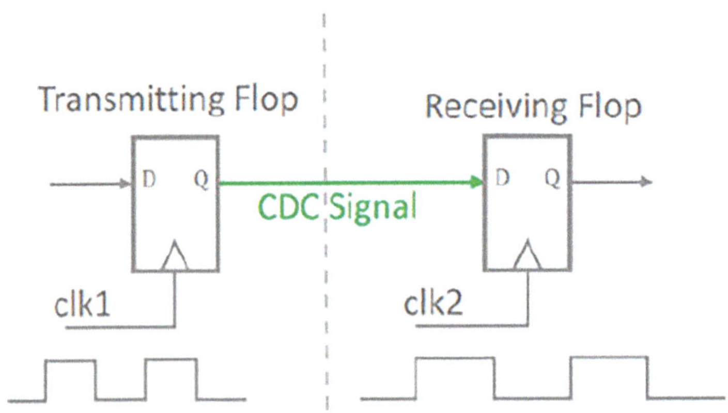

Fig. 1.10 CDC example

1.2.9 Simulation vs. Synthesis Mismatches

Some constructions behave differently in simulation and synthesis, which leads to false confidence from simulation-only verification.

1.2.9.1 Inferred Behavior Differences

In simulation, uninitialized memories may return 'X', while synthesis tools may implement them as 0.

1.2.9.2 Unintended Optimizations

RTL that is not written with synthesis in mind might get optimized out, changing functional behavior from simulation.

1.2.10 Tool-Related Hazards

Issues that arise not from the RTL code itself but from how tools interpret it or from version mismatches.

1.2.10.1 Lint and DRC Warnings Ignored

Lint tools often flag poor practices; ignoring them can hide bugs that appear much later in the design flow.

1.2.10.2 Inconsistent Tool Versions

Differences in synthesis or simulation tool versions (e.g., Synopsys DC vs. Cadence Genus) can lead to mismatched netlists and behaviors.

1.2.11 Loops

1.2.11.1 Combinational Feedback Loop

Combinational logic that feeds back into itself without passing through a sequential element (like a flip-flop or latch) can lead to unstable or oscillating outputs (Fig. 1.11). This feedback loop can cause **race conditions, non-deterministic behavior**, or even **combinational loops** that violate synthesis tool assumptions and lead to unpredictable timing or infinite simulation cycles. **Flip-flop does not, technically, have a feedback** (Fig. 1.12). Feedback is defined as an input extended from the output of the **SAME** device. But in case of Flip-flop, we extend the output to give it to the input of another gate of Flipflop. Break the combinational feedback loop as shown below by inserting FF (Figs. 1.13 and 1.14).

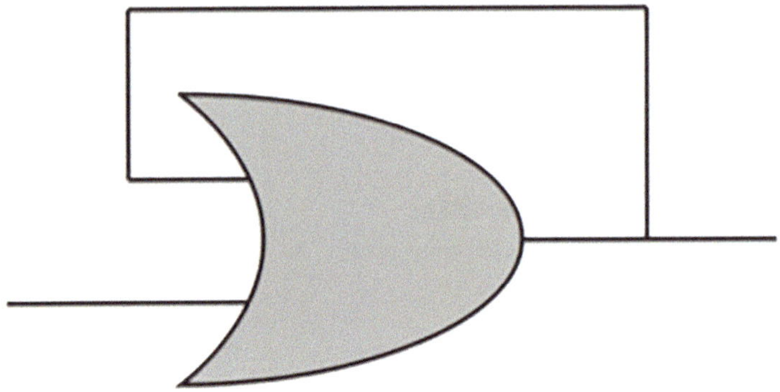

```
input a;
wire b;
 b = b | a;
```

Fig. 1.11 Combinational feedback loop

Fig. 1.12 FF

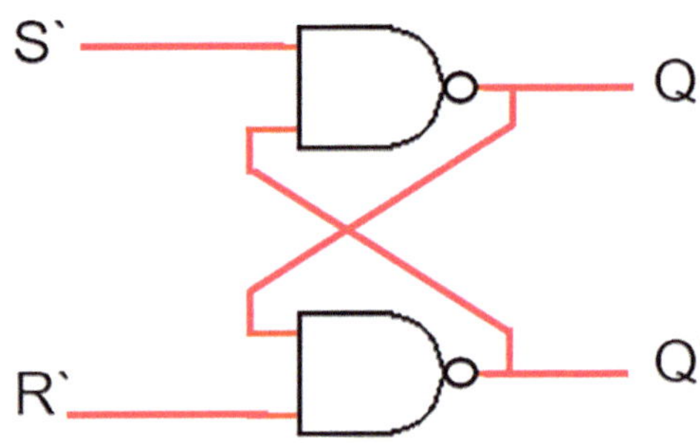

Fig. 1.13 The combinational feedback loop

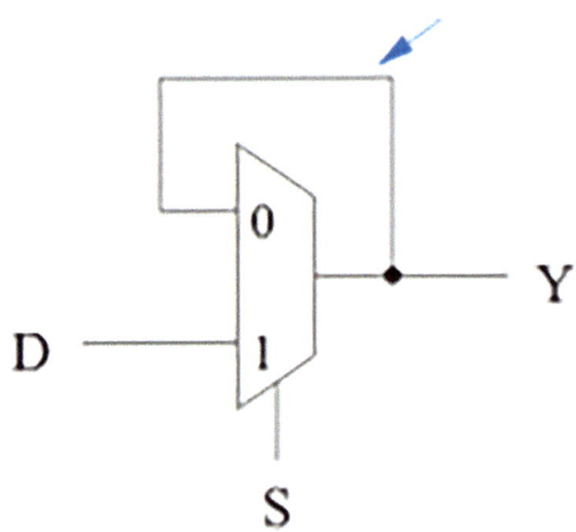

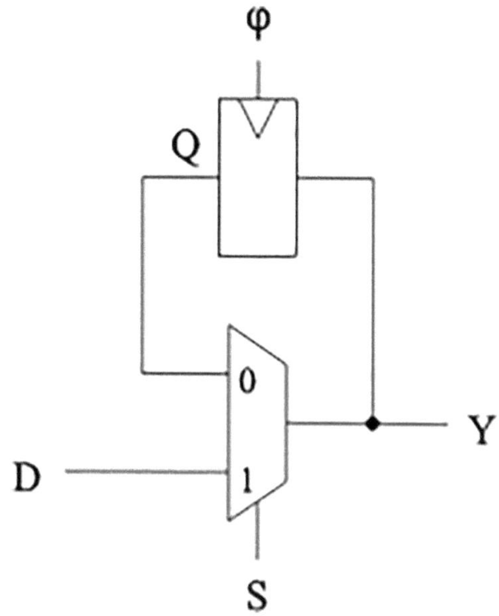

Fig. 1.14 Break the combinational feedback loop

1.2.11.2 Clock Loops

A clock loop is one that goes through CLK/Q path on a FF. This creates feedback into the clock network, leading to glitches. Use dedicated clock gating cells to avoid them (Fig. 1.15).

Glitch means combinational logic output changes many times when it is supposed to change once (Fig. 1.16). Glitch can cause hard-to-debug x propagation which lead to simulation mismatch with Synthesis. Glitches can happen too due to different delay paths.

1.2.12 Timing Constraints Mismanagement

Proper handling of timing constraints is crucial to ensure correct and efficient operation of the synthesized design. Misidentifying or failing to declare timing exceptions such as false paths and multi-cycle paths can mislead timing analysis tools, resulting in unreliable designs.

1.2.12.1 False Paths

A **false path** is a timing path that exists logically in the netlist but is never exercised functionally due to design constraints or control logic (Fig. 1.17). If not explicitly marked, synthesis and STA tools will attempt to meet timing on it unnecessarily, leading to **over-constraining**, area bloat, and longer runtimes.

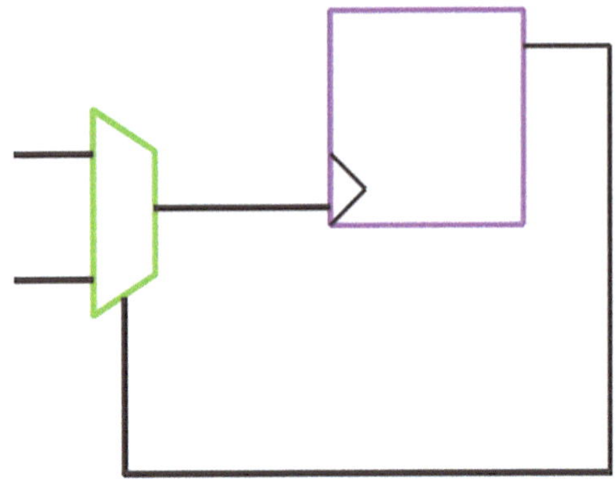

Fig. 1.15 Clock loop

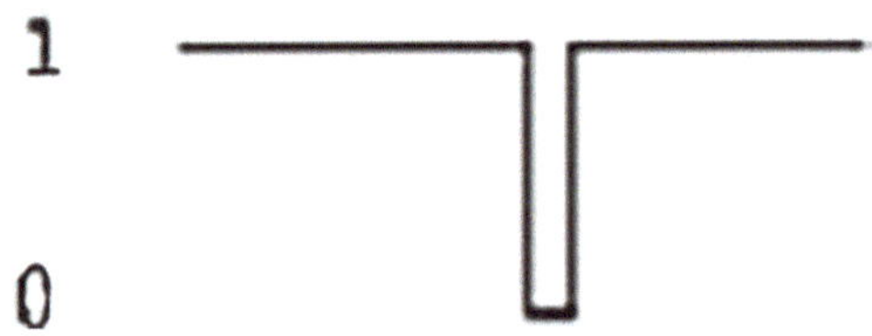

Fig. 1.16 A glitch example

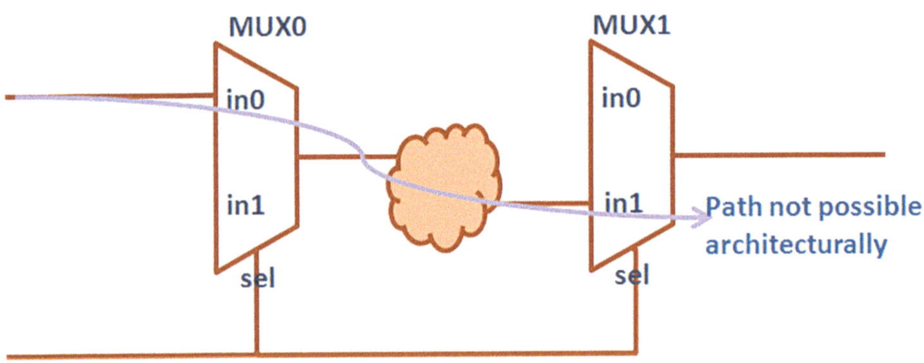

Fig. 1.17 False path

1.2.12.2 Multi-Cycle Paths

Multi-cycle paths are timing paths that are allowed to take more than one clock cycle to propagate data. This is common in slow-processing logic (e.g., a divider) or across coarse pipeline stages (Fig. 1.18).

1.2.12.3 Half-Cycle Paths

A half-cycle path is a timing path where the data is required to be captured within half a clock-cycle instead of the usual full clock period (Fig. 1.19). These paths are typically formed when:

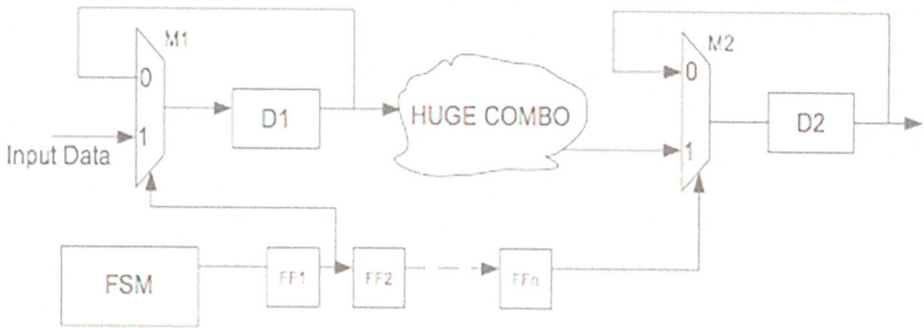

Fig. 1.18 Multi-cycle paths

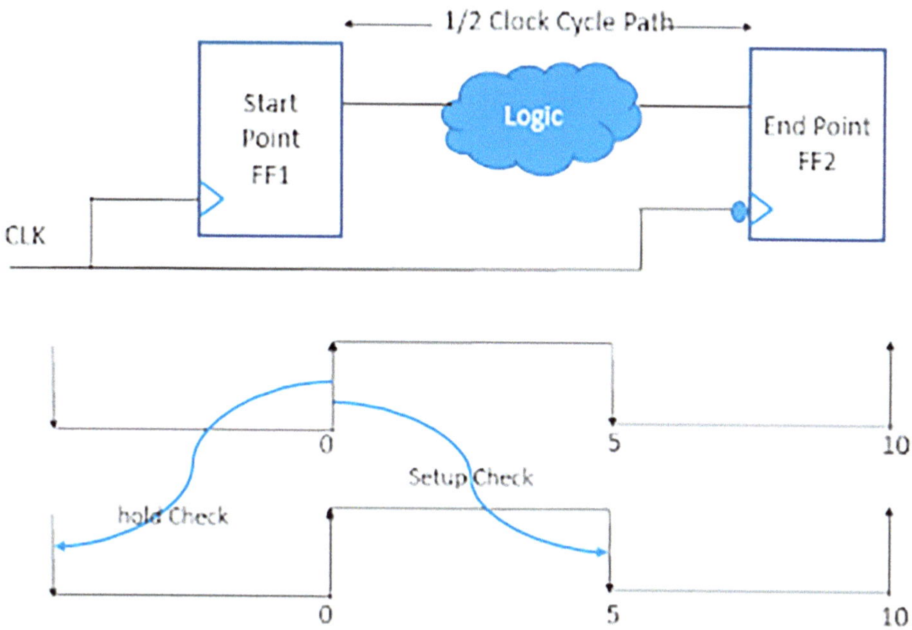

Fig. 1.19 A half-cycle path

- Data is launched on the rising edge of the clock and captured on the falling edge,
- Data is launched on the falling edge and captured on the rising edge.

This leads to tighter setup timing and relaxed hold timing:

- Setup timing becomes more stringent, as the data must be valid within half a clock cycle.
- Hold timing is more relaxed, as the capture register waits for the next opposite clock edge, effectively giving it 1.5 clock cycles to hold the data.

1.2.12.4 Setup and Hold Time Violations

Setup time is the minimum time interval before the active (rising or falling) clock edge during which the data input (D) must be stable and valid for the flip-flop to reliably capture the data. It is the window of time before the rising edge of the clock. Input must not change during this time.

Hold time is the minimum time interval after the active clock edge during which the data input (D) must remain stable and valid for the flip-flop to reliably capture the data. It is the window of time that is after the rising edge of the clock. Output must not change during this time.

Figure 1.20 shows the difference, where:

- Setup time (T_{su}): the minimum amount of time the data signal should be held steady before the clock.
- Hold time (T_h): the minimum amount of time the data signal should be held steady after the clock

Hold violations in a design occur when data changes too soon after a clock edge, violating the required **hold time**. Hold time refers to the minimum period that data must remain stable after the active clock edge to ensure it is correctly captured. If the data transitions during this period: typically, due to short paths or fast combinational logic: it may cause a hold violation.

Setup violations, on the other hand, arise when data arrives too late before the next clock edge, missing the required **setup time**. A common cause of setup violations is the presence of **long combinational paths or feedback loops**, which delay the data's arrival at the flip-flop, preventing it from being correctly latched in time.

To avoid **setup violations**, designers can apply **pipelining**, which involves breaking long combinational logic paths into shorter stages by inserting intermediate registers. This reduces propagation delay and ensures data reaches the destination flip-flop within the required setup time. Another common approach is to **lower the clock frequency**, effectively increasing the clock period and giving data more time to settle before the next edge.

To prevent **hold violations**, it's essential to **minimize clock skew** using **balanced clock tree designs**, ensuring clock signals arrive at flip-flops simultaneously. Additionally, **inserting delay buffers** in short data paths can help keep the data stable long enough after the clock edge, satisfying the required hold time.

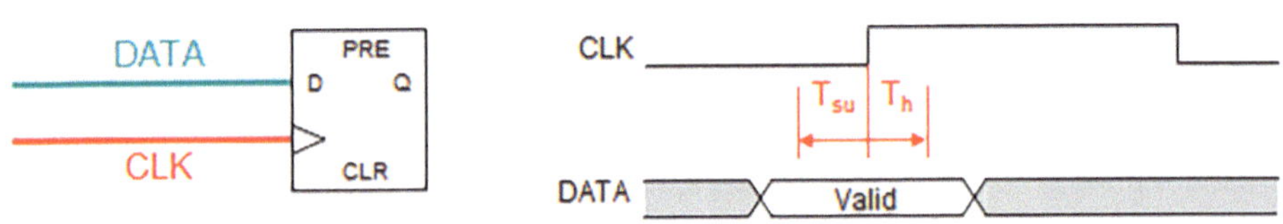

Fig. 1.20 Setup and hold time

1.2.13 Security Vulnerabilities in RTL Design

Modern RTL designs, especially in processors and accelerators, are susceptible to security threats such as transient execution side channels and hardware Trojans. These vulnerabilities can be exploited to leak sensitive information or compromise system integrity. Traditional verification methods may not detect these issues, necessitating formal verification techniques like Unique Program Execution Checking (UPEC) to ensure security properties are upheld [12].

1.3 Improving Design Process for RTL Design

Improving the RTL design process is essential to simplify and accelerate debugging, which remains one of the most time-consuming phases of digital design development [13–19]. Tables 1.2 and 1.3 summarize all these techniques.

1.3.1 Optimizing Data Paths and Logic

1.3.1.1 Reduce Logic Depth

Reducing the depth of combinational logic is an essential technique for improving FPGA design performance. Deep logic paths, which occur when multiple logic operations are chained together in a single combinational block, can significantly increase the overall delay and reduce the clock frequency. By breaking complex logic into smaller, manageable stages, the overall depth is minimized, making the design faster and more efficient. This method helps mitigate long propagation delays, thus optimizing the timing of the design.

1.3.1.2 Pipeline the Design

Pipeline design is a common and effective technique for improving performance in FPGA designs. By introducing pipeline stages, long data paths are divided into shorter segments, reducing the critical path and allowing higher clock frequencies. Registers placed between stages help in separating the computational stages, allowing for faster processing of data and improving throughput. Pipelines are especially beneficial in designs with repetitive computations or high data throughput requirements, where latency reduction is crucial.

1.3.1.3 Minimize Fanout

Fanout refers to the number of destinations to which a signal is connected. High fanout can cause routing congestion and lead to longer delays in the FPGA design. To minimize fanout, engineers can insert buffers or registers that distribute the signal to multiple destinations. This reduces the load on any single wire or routing path, improving the overall

Table 1.2 Design-dependent solutions

Category	Technique	Description
Code efficiency: Optimizing data paths and logic	Reduce Logic Depth	Break up complex logic into smaller, more manageable stages to minimize the depth of combinational logic. Avoid deep nesting of logic elements.
	Pipeline the Design	Introduce pipeline stages to divide long data paths into shorter segments, reducing the critical path and improving the clock frequency. Use registers to separate computational stages.
	Minimize Fanout	Reduce fanout by inserting buffers or registers to distribute signals to multiple destinations, reducing routing congestion.
	Avoid Latches	Avoid the use of latches in the design, as they can complicate routing due to the required feedback path for the latched signal.
	Floor Planning	Carefully consider floor planning to place critical components closer to each other, reducing the length of critical connections.
	Avoid Complex Conditional Statements	Minimize the complexity of conditional statements and loops. Overly complex control logic can increase routing congestion.
	Loop Unrolling	In cases where loops are performance-critical, consider loop unrolling to reduce control logic and optimize routing.
	Optimize code structure	Write clean and efficient code with proper hierarchical design. Use modules and functions to encapsulate logical blocks, improving readability and reusability.
	Use proper data types	Choose appropriate data types that match the intended purpose of variables and signals. Using the smallest possible data type can reduce resource usage.
	Limit design hierarchy levels	Avoid deep levels of hierarchy as it can increase the synthesis time.
	Optimize clock and reset logic	Carefully design clock and reset logic to minimize the number of clock domains and ensure synchronous and reliable operation.
	Use appropriate coding styles	Follow coding guidelines and use Verilog constructs that are more synthesis-friendly.
	Analyze and reduce unnecessary operations	Perform code analysis to identify any unnecessary operations or redundant logic. Eliminate unused signals, dead code, or redundant computations to optimize the design.

(continued)

Table 1.2 (continued)

Category	Technique	Description
Clock and synchronization	Clock Domain Management	Clearly define and manage clock domains in your design. Use synchronous design principles to minimize metastability issues when crossing clock domains.
	Utilize Multiclock Domains	Take advantage of FPGA tools' support for multiple clock domains to reduce the impact of routing and synchronization between different clock regions.
Memory and Resource optimization	Use FPGA-Specific Primitives	Utilize FPGA-specific resources and primitives, such as block RAM and DSP slices, for tasks that can benefit from these resources. Take advantage of built-in functions provided by FPGA toolchains.
	Resource Sharing	Minimize resource usage by sharing resources where possible.
	Optimize for RAM and ROM Usage	Optimize memory usage by reducing unnecessary read and write operations. Use efficient memory access patterns to avoid creating memory bottlenecks.
Timing and Constraints	Timing Constraints	Set accurate timing constraints to guide the synthesis and place-and-route tools.

Table 1.3 Tool-dependent solutions

Technique	Description
Synthesis and Optimization Tools	Experiment with different synthesis and optimization settings in your FPGA design tool to see how they impact routing and performance.
Tool Vendor Recommendations	Follow guidelines and recommendations provided by the FPGA tool vendor. They may have specific strategies for optimizing routing.
Utilize Synthesis Directives and Pragmas	Take advantage of synthesis directives provided by your synthesis tool. These directives allow you to guide the synthesis process and optimize specific portions of your code.

efficiency of the design and decreasing the risk of timing violations due to congested routing.

1.3.1.4 Avoid Latches

While latches can be used to store data, they introduce complexities in FPGA designs. Latches require feedback paths, which can complicate routing and increase the overall design's critical path. These feedback paths can also make timing analysis more difficult. As a result, it is recommended to avoid using latches in designs when possible, opting for flip-flops or other sequential elements that do not require feedback paths, ensuring a more straightforward and efficient routing process.

1.3.1.5 Floor Planning

Effective floor planning is crucial for optimizing FPGA designs. By carefully placing critical components close to one another, the length of the critical connections is reduced, leading to improved performance. Shorter routing paths allow for faster signal propagation, which directly impacts the clock speed and the overall efficiency of the design. A well-organized floor plan also makes it easier to manage the routing of signals, minimizing congestion and improving the overall synthesis results.

1.3.1.6 Avoid Complex Conditional Statements

Complex conditional statements and loops can increase the overall complexity of the control logic, leading to higher resource usage and routing congestion. Simplifying these statements or breaking them down into more manageable parts can reduce the amount of logic and enhance the design's efficiency. Additionally, minimizing complex conditions helps to avoid introducing long control paths, which can negatively impact the timing and performance of the design.

1.3.1.7 Loop Unrolling

Loop unrolling is a technique used to optimize performance in FPGA designs that include loops. By unrolling loops, multiple iterations of the loop are executed in parallel, reducing the overhead of control logic and optimizing routing. This is particularly useful when loops are performance-critical and need to be executed as fast as possible. However, this technique increases resource usage as multiple copies of the loop logic are created, so it should be used judiciously.

1.3.2 Optimize Code Structure

A clean, hierarchical code structure is vital for efficient FPGA design. By using modules and functions to encapsulate logical blocks, the design becomes more readable, reusable, and easier to debug. A well-structured design also aids the synthesis process by providing clear separation of concerns, allowing the tool to optimize individual blocks more effectively. Additionally, a proper hierarchical structure helps in managing complexity, especially in large designs.

1.3.2.1 Use Proper Data Types

Selecting the appropriate data types for variables and signals can have a significant impact on resource usage and performance. Using the smallest possible data type for a variable can reduce the number of resources needed to implement the design, especially in large designs with many signals. Using the correct data type also helps ensure that the FPGA can handle the design more efficiently, leading to better optimization and faster processing.

1.3.2.2 Limit Design Hierarchy Levels

Deep levels of design hierarchy can lead to increased synthesis time and slower overall design flow. Limiting the number of hierarchy levels helps streamline the synthesis process, reducing unnecessary complexity and improving the tool's ability to optimize the design. A flat structure or shallow hierarchy can often lead to faster implementation and synthesis times, which is especially important in time-sensitive design projects.

1.3.2.3 Optimize Clock and Reset Logic

Clock and reset logic play a crucial role in the design of FPGA circuits. By minimizing the number of clock domains and carefully designing the reset logic, you can ensure that the design operates synchronously and reliably. Synchronizing all clock domains reduces the risk of metastability, ensuring that signals are stable when crossing boundaries between different clock regions. Optimizing these aspects helps avoid unnecessary clocking resources and improves the overall stability and reliability of the design.

1.3.2.4 Use Appropriate Coding Styles

Using coding styles that are optimized for synthesis is crucial for improving the efficiency of FPGA designs. Certain Verilog constructs are more synthesis-friendly than others, and by adhering to coding guidelines, designers can avoid inefficient patterns that may lead to excessive resource usage or longer synthesis times. Proper coding practices can also help the synthesis tool understand the designer's intentions, leading to better optimizations.

1.3.2.5 Analyze and Reduce Unnecessary Operations

Unnecessary operations or redundant logic can bloat an FPGA design, increase resource usage and slowing down synthesis times. Performing a thorough code analysis helps identify and eliminate unused signals, dead code, or redundant calculations. By removing these inefficiencies, the design becomes more streamlined and efficient, leading to faster synthesis and improved performance.

1.3.3 Clock and Synchronization

1.3.3.1 Clock Domain Management

Managing clock domains is critical for maintaining the integrity and stability of an FPGA design. When working with multiple clock domains, it's essential to carefully define and manage the interactions between them to avoid issues like metastability or data corruption. Employing synchronous design principles ensures that data crossing clock boundaries remains stable and reliable, improving the overall timing of the design.

1.3.3.2 Utilize Multiclock Domains

FPGA tools support the use of multiple clock domains within a design, which can be beneficial for managing complex systems with different timing requirements. By taking advantage of multi-clock domains, you can reduce the impact of routing congestion and timing issues between different clock regions. This approach allows the design to scale effectively without compromising performance or routing efficiency.

1.3.4 Memory and Resource Optimization

1.3.4.1 Optimize for RAM and ROM Usage

Efficient use of RAM and ROM is crucial in optimizing FPGA designs. Reducing unnecessary read and write operations, as well as optimizing memory access patterns, can help avoid memory bottlenecks. Proper memory management ensures that the design makes efficient use of available memory resources, leading to better performance and lower resource consumption.

1.3.5 Timing and Constraints

1.3.5.1 Timing Constraints

Accurate timing constraints are essential for guiding the synthesis and place-and-route tools in FPGA design. By setting precise constraints, you can ensure that the design meets timing requirements, and the tool can optimize the placement of logic elements for the best performance. Timing constraints help manage delays and ensure that the design operates within the required frequency, preventing timing violations and ensuring reliable operation.

References

1. Fang, W., Yao, L., Liu, S., Zhang, Q., Ceyu, X., Wills, L. W., Zhang, H., & Xie, Z. (2023). MasterRTL: A pre-synthesis PPA estimation framework for any RTL design. In *2023 IEEE/ACM international conference on computer aided design (ICCAD)* (pp. 1–9). IEEE.
2. Salah, K. (2017). MCERTL: Mutation-based correction engine for register-transfer level designs. *World Academy of Science, Engineering and Technology. International Journal of Electronics and Communication Engineering. International Scholarly and Scientific Research & Innovation, 11*(11) https://publications.waset.org/abstracts/80068.pdf
3. Fadiheh, M. R., Wezel, A., Müller, J., Bormann, J., Ray, S., Fung, J. M., Mitra, S., Stoffel, D., & Kunz, W. (2021). An exhaustive approach to detecting transient execution side channels in RTL designs of processors. *arXiv:2108.01979.*
4. W. Fang, Y. Lu, S. Liu, Q. Zhang, C. Xu, L. W. Wills, H. Zhang and Z. Xie, "MasterRTL: A pre-synthesis PPA estimation framework for any RTL design," in Proc. IEEE/ACM int. conf. computer-aided design (ICCAD), IEEE 2023, pp. 1–9.
5. Chattopadhyay, A., & Sutrakar, V. K. (2025). Machine learning framework for early power, performance, and area estimation of RTL. *arXiv:2502.16203.*
6. Abdelatty, M., Ma, J., & Reda, S. (2025). MetRex: A benchmark for Verilog code metric reasoning using LLMs. In *ASP-DAC 2025, Providence, RI*. ACM.

7. Thakur, S., et al. (2022). Benchmarking large language models for automated Verilog RTL code generation. *arXiv:2212.11140.*
8. Abdelatty, M., Ma, J., & Reda, S. (2022). How good is your RTL code? A quick answer from machine learning. In *IEEE/ACM ICCAD*. IEEE.
9. Mohamed, K. S. (2016). IP cores design from specifications to production: Modeling, verification, optimization, and protection. In *IP cores design from specifications to production: Modeling, verification, optimization, and protection*. Springer International Publishing.
10. Mohamed, K. S. (2024). An introduction to heterogeneous SoC design and verification: A conceptual-level. In *Heterogeneous SoC design and verification. Synthesis lectures on digital circuits and systems*. Springer.
11. Davidmann, S. J., & Flake, P. L. (2006). *SystemVerilog for design: A guide to using SystemVerilog for hardware design and modeling*. Springer.
12. Fadiheh, M. R., Wezel, A., Mueller, J., Bormann, J., Ray, S., Fung, J. M., Mitra, S., Stoffel, D., & Kunz, W. (2021). An exhaustive approach to detecting transient execution side channels in RTL designs of processors. *arXiv:2108.01979.*
13. Pei, Z., Zhen, H.-L., Yuan, M., Huang, Y., & Yu, B. (2024). BetterV: Controlled Verilog generation with discriminative guidance. *arXiv:2402.03375.*
14. Zhao, Z., Qiu, R., Lin, I.-C., Zhang, G. L., Li, B., & Schlichtmann, U. (2025). VRank: Enhancing Verilog code generation from large language models via self-consistency. *arXiv:2502.00028.*
15. Thakur, S., et al. (2023). VeriGen: A large language model for verilog code generation. *arXiv:2308.00708v1.*
16. Liu, M., et al. (2023). VerilogEval: Evaluating large language models for Verilog code generation. *arXiv:2309.07544.*
17. Thakur, S., et al. (2022). Benchmarking large language models for automated Verilog RTL generation. *arXiv:2212.11140.*
18. Yang, G., et al. (2024). Large language model for Verilog generation with golden code feedback. *arXiv:2512.00020.*
19. Nadimi, B., et al. (2024). A multi-expert large language model architecture for Verilog code generation. *arXiv:2404.08029.*

Introduction: Basics of RTL Debugging 2

2.1 Problem-Solving Strategies vs Debugging Strategies

Problem-solving strategies and debugging strategies share a common goal of resolving issues, but they differ significantly in their approach, tools, mindset, and techniques. Problem-solving strategies are aimed at finding and implementing solutions to known or emerging issues. They typically begin with a recognized problem or a need for improvement, requiring a creative, proactive mindset. Problem-solving involves either top-down or bottom-up analysis, where the issue is approached by understanding the system as a whole or breaking it down into smaller components. Techniques such as root cause analysis, system decomposition, and hypothesis testing are often employed, with tools like whiteboards, documentation, models, and design specifications used to support the process. The focus is on understanding the "why" of the problem and finding a way forward, such as designing a new interconnect to improve performance (Table 2.1).

On the other hand, debugging strategies are focused on identifying, isolating, and fixing bugs in a design or implementation. The process begins with an observed failure, mismatch, or unexpected behavior, which requires a systematic and investigative approach. Debugging often involves step-by-step tracing of code execution, examining waveforms, logs, and assertions to uncover the root cause of the issue. The main focus is on understanding "how" the failure occurred and restoring the correct behavior, typically through techniques like signal probing, simulation, assertions, and protocol checks. Tools commonly used in debugging include simulators, waveform viewers, debug logs, and protocol analyzers. The mindset is reactive, as the primary goal is to respond to and resolve a malfunction. An example of debugging would be troubleshooting a deadlock in a finite state machine (FSM) after simulation shows a system hang.

K. Mohammed, *The Art of RTL Debugging*, Synthesis Lectures on Digital Circuits & Systems, https://doi.org/10.1007/978-3-032-15880-2_2

Table 2.1 Problem-solving strategies vs debugging strategies

Aspect	Problem-solving strategies	Debugging strategies
Goal	To find and implement a solution to a known or emerging issue	To identify, isolate, and fix a design or implementation bug
Starting point	A known problem or need for improvement	An observed failure, mismatch, or unexpected behavior
Approach	Top-down or bottom-up analysis, brainstorming, modeling	Step-by-step tracing, waveforms, logs, assertions
Focus	Understanding the why and finding a path forward	Understanding the how it failed and restoring correct behavior
Techniques used	Root cause analysis, system decomposition, hypothesis testing	Signal probing, simulation, assertions, protocol checks
Common tools	Whiteboard, documentation, models, design specs	Simulators, waveform viewers, debug logs, protocol analyzers
Mindset	Creative, explorative, often proactive	Investigative, systematic, often reactive
Example	Designing a new interconnect to improve performance	Debugging a deadlock in an FSM after simulation shows a hang

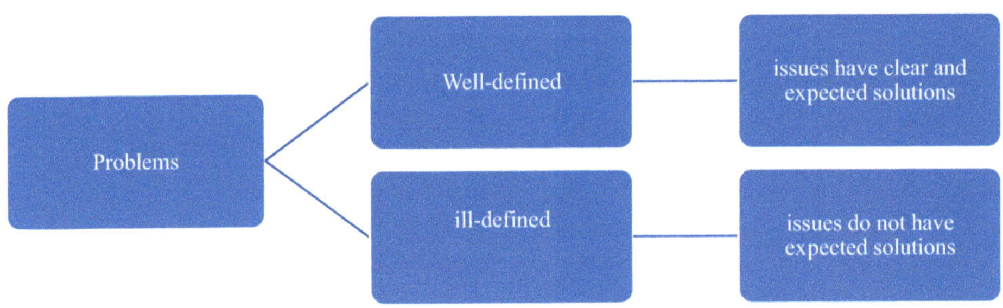

Fig. 2.1 Problem classifications

While problem-solving strategies are more about creative thinking and designing solutions from the ground up, debugging strategies are a methodical, reactive process of investigating failures and restoring functionality [1–3].

2.2 Problem Solving Strategies: Algorithmic Paradigms

The process of problem-solving involves the classification of problems into two categories: ill-defined and well-defined problems, as depicted in Fig. 2.1. Before devising a solution, it is imperative to accurately identify the problem at hand. Various strategies for problem-solving are outlined in Table 2.2 [4]. Algorithms, comprising a set of instructions for specific tasks, play a crucial role in processing vast amounts of data, particularly in the context of big data. The problem-solving process entails steps such as problem identification, comprehension, exploration of alternative solutions, selection of the optimal approach, and evaluation of the solution's efficacy. Time complexity, illustrated in Table 2.3, defines the order of computation time for an algorithm. Algorithms with

Table 2.2 Problem solving strategies

Strategy	Explanation
Trial and error	Continue trying different solutions until problem is solved
Algorithm	Step-by-step problem-solving formula
Heuristic	General problem-solving framework based on past experiences

polynomial time complexity $\mathbf{O(n^2)}$ are considered manageable, whereas those with non-polynomial (NP) time complexity present more formidable challenges. Problems exhibiting NP time complexity are termed NP-Hard Problems, which can be addressed using machine learning (ML) techniques to resolve long-standing circuit optimization challenges. Performance evaluation of algorithms relies on Asymptotic Analysis, where the algorithm's efficiency is gauged in relation to input size without measuring actual running time. Each problem represents a feature space encompassing all potential solutions, necessitating efficient search strategies for resolution. Expert knowledge, specific to domains like medicine, finance, or engineering, aids in problem-solving and is referred to as the knowledge domain. The problem domain, depicted in Fig. 2.2, encompasses all potential problem areas and is inherently broader than any individual knowledge domain [5–10].

Algorithmic paradigms can be classified as:

- Brute Force
- Greedy Algorithm
- Linear Programing
- Dynamic Programing: Divide and conquer/smart recursion
- Machine Learning: to solve NP-Complete Problems
- Heuristic algorithm: random search or stochastic
- Meta-Heuristics Algorithm: such as genetic algorithm
- Randomness and Probability
- Dual Hill Climbing: Max Flow/Min Cut
- Recursive Back Tracking: Bellman-Ford
- First Order Logic
- Graph Algorithms: DFS/BFS/Dijkstra's

2.2.1 Dynamic Programming

Dynamic programming (DP) breaks a problem into smaller sub-problems and then uses some smart recursion such that complexity of solution can come down from exponential complexity to polynomial complexity. Smart recursion means it uses a table to store the results of sub-problem so that if same sub-problem is encountered again in future, it could

Table 2.3 Classification of time complexity [1]

Notation	Complexity	Description	Example	

O(1)	Constant	Simple statement	Addition	`int a = 5, b = 10, sum;`
				`sum = a + b;`
				`printf("%d", sum);`

O(log(n))	Logarithmic	Divide in half	Binary search	`Statement 1`
				`Statement 2`
				.
				.
				.
				`Statement k`

O(n)	Linear	Loop	Linear search	`int i = 1;`
				`while(i < 10) {`
				`    printf("%d", i);`
				`    i = i * 2;`
				`}`

O(n*log(n))	Linearithmic	Divide and conquer	Merge sort	`for (i = 1; i < n; i++) {`
				`    printf("%d", i);`
				`}`

$O(n^2)$	Quadratic	Double loop	Check all pairs	`for (int x = 0; x < n; x++) {`
				`    for (int y = 0; y < n; y++) {`
				`        sum += x * y;`
				`    }`
				`}`

$O(n^3)$	Cubic	Triple loop	Check all triples	`for (int a = 0; a < n; a++) {`
				`    for (int b = 0; b < n; b++) {`
				`        for (int c = 0; c < n; c++) {`
				`            sum += a * b * c;`
				`        }`
				`    }`
				`}`

$O(2^n)$	Exponential	Exhaustive search	Check all subsets	`decimal Fibonacci(int n) {`
				`    if (n == 0)`
				`        return 0;`
				`    else if (n == 1)`
				`        return 1;`
				`    else`
				`        return Fibonacci(n - 1) + Fibonacci(n - 2);`
				`}`

$O(n!)$	Factorial	Recursive function	Factorial	`int factorial(int n) {`
				`    if (n == 0)`
				`        return 1;`
				`    else`
				`        return n * factorial(n - 1);`
				`}`

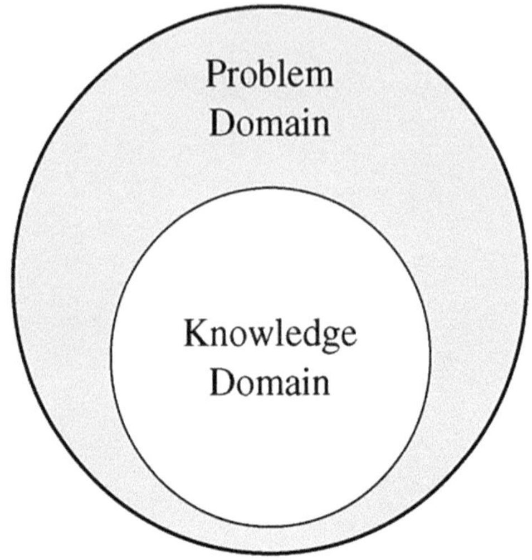

Fig. 2.2 Problem domain vs Knowledge domain

directly return the result instead of re-calculating it (Fig. 2.3). It is like divide and conquer. A DP problem might have linear or non-linear cost function and constraints. Dynamic programming treats time as a necessary step in scheduling.

2.2.2 Linear Programming

Linear programming (LP) is an attempt to find a maximum or minimum solution to a function, given certain constraints where the optimization cost functions and constraints are linear i.e. they are always a line not curved or in any weird shapes (Fig. 2.4). A linear programming problem can be solved using DP if the problem is broken into smaller sub-problems and followed DP approach. Linear programming is a subset of dynamic programming.

2.2.3 Greedy Algorithm

Greedy algorithm finds the best solution to a problem one step at a time. At each step, the algorithm makes the choice that improves the solution the most, even if that choice makes future steps less fruitful. It always makes a choice that seems best "right now", without considering the future implications of this choice. A binary tree can be used to represent greedy algorithm problems. In this binary tree representation, each node corresponds to a decision point in the greedy process, and each branch represents a possible choice made at that step. As the algorithm progresses, it traverses the path that yields the locally optimal solution at each level of the tree. While this approach does not always lead to the globally optimal solution, it offers a computationally efficient way to approach complex problems,

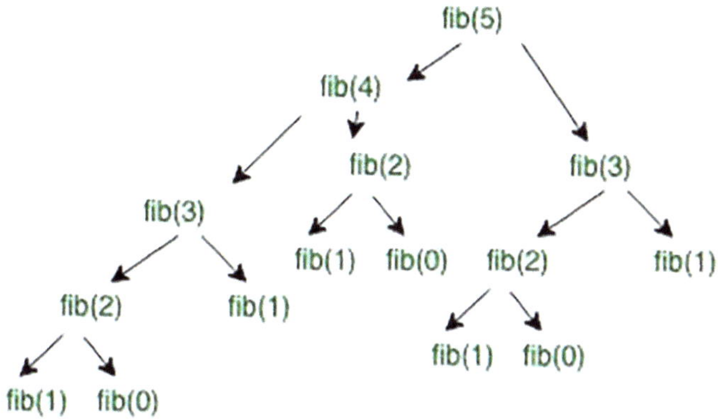


```
def fib(n):
    if n <= 1:
        return n

    return fib(n - 1) + fib(n - 2)
```

Fig. 2.3 Dynamic Programming: a simple recursive program for Fibonacci numbers explained in Python and in a binary tree representation

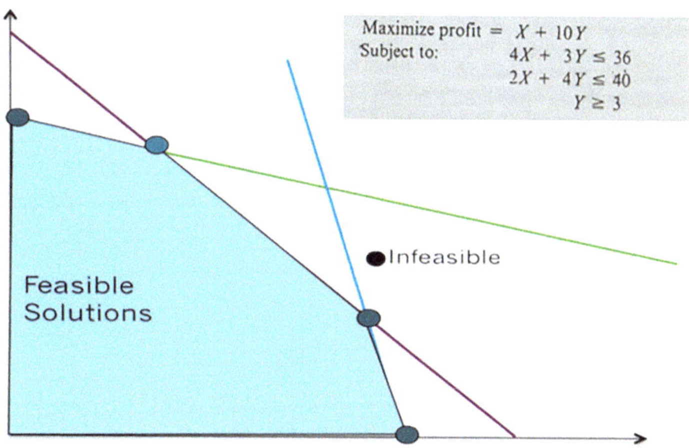

Fig. 2.4 Concept of linear programming: a graphical explanation

especially when the problem exhibits the **greedy-choice property** (a globally optimal solution can be arrived at by making a series of locally optimal choices) and **optimal substructure** (an optimal solution to the problem contains optimal solutions to its subproblems). The binary tree structure also helps visualize alternative paths that could have been taken, which is particularly useful when analyzing why the greedy algorithm might succeed or fail in a given context.

Common examples of problems well-suited for greedy algorithms include **Huffman coding**, **Prim's algorithm** for minimum spanning trees, and **Dijkstra's algorithm** for shortest paths (when all edge weights are non-negative). In such cases, the greedy decisions at each node of the binary tree can efficiently converge to an optimal solution, validating the effectiveness of the approach.

2.2.4 Brute-Force Algorithm

Brute-Force algorithm approach is to try all possible patterns to find an answer. It describes an algorithm that checks every possible scenario rather than using some type of trick to speed the process up. This approach guarantees correctness by exhaustively evaluating every potential solution, making it a reliable baseline for solving problems—especially when correctness is more important than efficiency. However, brute-force algorithms tend to be computationally expensive, as their time complexity often grows exponentially or factorially with input size. As a result, they are typically only feasible for small input domains or for problems where all other heuristic or optimized methods have failed [11].

A brute-force search can be visualized using structures such as **trees or graphs**, where each node represents a partial solution and each branch extends toward a possible continuation. The algorithm systematically explores all branches until a valid solution is found or the search space is exhausted. Although not efficient, this method has the advantage of being simple to implement and free from problem-specific assumptions.

Common examples of brute-force applications include **password cracking by trying all character combinations**, **solving puzzles like Sudoku through exhaustive trial**, or **string pattern matching using naive search**, where every character in the text is checked against every possible starting point of the pattern. Despite its inefficiency, brute-force remains an important concept in algorithm design, both as a fallback and as a foundation for understanding more advanced optimization techniques such as **backtracking**, **dynamic programming**, or **greedy algorithms**.

2.2.5 Heuristic Algorithm

Heuristic algorithm finds solutions among all possible ones without guaranteeing that the best will be found, therefore they may be considered as approximate and not accurate algorithms. Heuristics are designed to provide **good-enough solutions** when exhaustive search is impractical, especially for problems with very large search spaces or when rapid responses are needed. These algorithms deliberately leverage domain knowledge, rules of thumb, or problem-specific insights to guide the search toward promising regions, without requiring an exhaustive check of every possibility. As a result, they often achieve substantial reductions in computation time while yielding reasonably high-quality solutions.

Heuristic algorithms are typically visualized using **search spaces**, **graphs**, or **state trees**, where they selectively explore only the most promising branches based on evaluation metrics or scoring functions rather than traversing everything. Because they do not ensure global optimality, their output is considered **approximate**. However, when the heuristic is well-designed, the solution is often sufficiently close to optimal for practical purposes.

Classic examples include **best-first search** and **A*** (A-star) for pathfinding, which prioritize nodes based on heuristic estimates of distance to the goal; **simulated annealing**, which probabilistically accepts worse moves early on to escape local minima; and **genetic algorithms**, which evolve candidate solutions via selection, crossover, and mutation based on fitness evaluation. In these cases, the heuristic strategy dramatically reduces search overhead while still uncovering competitive solutions. They are especially useful in complex domains such as scheduling, routing, optimization under constraints, and real-time decision making in AI [12].

2.2.6 Meta-Heuristics Algorithm

Meta-Heuristics Algorithm means it is neither fully stochastic nor fully deterministic, but a mix of both. An example is genetic algorithm where the part of fitness function is deterministic, but crossover and mutation parts are random. Meta-heuristic algorithms are higher-level problem-solving frameworks designed to guide subordinate heuristics in exploring large and complex search spaces efficiently. Unlike simple heuristics, metaheuristics incorporate both **deterministic logic** (structured decision-making) and **stochastic elements** (randomized behaviors) to balance exploration (trying diverse solutions) and exploitation (refining known good solutions). This hybrid nature allows them to avoid local optima and adapt to a wide variety of optimization problems without needing a complete understanding of the problem's structure.

They are especially useful when the search space is rugged, high-dimensional, or discontinuous, where deterministic algorithms alone may struggle to escape suboptimal traps. The deterministic part often governs rules like selection criteria or stopping conditions, while the stochastic components introduce randomness to promote diversity in the solution pool or to enable occasional jumps to new areas of the search space [13].

In addition to **Genetic Algorithms (GA)**, which use deterministic fitness evaluation and randomized genetic operators, other well-known meta-heuristics include **Simulated Annealing** (which uses probabilistic acceptance of worse solutions to escape local minima), **Particle Swarm Optimization (PSO)** (where particles follow deterministic motion influenced by randomness), and **Ant Colony Optimization (ACO)** (where decisions are probabilistically made based on pheromone trails and heuristics). These algorithms are widely applied in areas like scheduling, routing, machine learning model tuning, and real-time control systems—where exhaustive or purely heuristic methods would be infeasible.

Their versatility lies in the fact that they do not guarantee optimality, but are capable of finding **near-optimal solutions** with reasonable computational effort, especially for **NP-hard** and **combinatorial optimization** problems.

2.3 Workaround vs Fixing in RTL Debug

A **workaround** involves implementing a solution that circumvents the underlying issue without addressing the root cause of the problem. It is a temporary or partial solution designed to keep the system functional while the actual issue is investigated further or until a more permanent fix can be made (Table 2.4).

When to Use a Workaround:

- **Time Constraints:** If there's limited time to meet project deadlines and a quick solution is needed to proceed.
- **Non-Critical Issues:** When the issue does not affect the primary functionality of the design or occurs in a non-critical section.
- **Isolating Complex Problems:** When debugging an issue is too complex or time-consuming, a workaround can help maintain progress until the design team can address the root cause.

Example Workaround:

- **Adding Redundant Logic:** If a race condition is detected but difficult to fix immediately, adding a temporary synchronization flip-flop to ensure signal stability might be considered a workaround.
- **Using a Clock Divider:** If a clock skew issue arises, using a clock divider to slow down the clock might be used as a workaround to give timing issues more margin, rather than fixing the underlying clock synchronization problem.

Table 2.4 Workaround vs fixing

Aspect	Workaround	Fixing
Time investment	Quick, temporary solution	Time-consuming, permanent solution
Impact on design	May impact performance or add complexity	Improves robustness and design quality
Risk of new bugs	Lower risk of introducing new bugs	Higher risk due to changes made to core logic
When to use	When facing time constraints, non-critical issues	When the issue affects functionality or quality
Long-term outcome	Temporary relief, might not be scalable	Permanent resolution, enhances stability

Pros of Workaround:

- **Quick Solution:** Helps to meet project deadlines or milestones.
- **Low Risk:** In some cases, workarounds don't modify the core logic and minimize risk of new issues.

Cons of Workaround:

- **Temporary Solution:** It doesn't address the root cause of the issue.
- **Can Mask the Real Problem:** Workarounds may hide the true design flaw, leading to future issues.
- **Potential Performance Impact:** Some workarounds, like adding redundant logic, may affect the overall performance of the design.

Fixing refers to resolving the root cause of a problem by identifying the source of the issue and making the necessary changes to the design. This typically involves modifying the RTL code, improving timing constraints, or adjusting design methodology to permanently eliminate the hazard.

When to Fix the Issue:

- **Critical Bugs:** When the issue affects core functionality or causes significant operational problems.
- **Long-Term Stability:** When the issue is likely to affect future versions of the design or cause long-term problems if not addressed properly.
- **Regulatory or Quality Standards:** When a formal review, such as from a customer or for certification, requires that all issues be fully resolved.

Example Fix:

- **Race Condition Resolution:** Identifying the exact location where signals are improperly synchronized and adding a proper synchronization flip-flop or changing the clocking scheme to fix the race condition.
- **FSM Dead States:** Correcting an FSM that contains unused or unreachable states by eliminating these states or adding necessary transitions to ensure proper operation.

Pros of Fixing:

- **Long-Term Solution:** Resolves the underlying issue and prevents future recurrence.
- **Improves Design Quality:** Leads to a more robust and reliable design.
- **Meets High-Quality Standards:** Often required for meeting industry standards and certifications.

Cons of Fixing:

- **Time-Consuming:** Can be more time-intensive, especially if the root cause is complex or hard to identify.
- **Risk of New Bugs:** Changes to the design might introduce new issues or require other parts of the design to be revalidated.

A workaround and a proper bug fix differ fundamentally in intent and depth. Fixing a bug means identifying the root cause of a problem and correcting it so that the issue is resolved at its source. A workaround, on the other hand, is more like sidestepping the issue: it doesn't eliminate the underlying problem but instead finds a way to avoid triggering it or to lessen its impact. In some cases, a workaround can feel like you're simply changing the name of the problem rather than solving it. It's as if you're putting a label over a crack in the wall instead of repairing the structure itself. While workarounds can be useful for short-term relief or when time constraints exist, they are not substitutes for real fixes when long-term stability and correctness are the goals.

2.4 History of Debugging

The history of debugging dates back to the early days of computer programming. The term "**bug**" in the context of programming is often attributed to Grace Hopper, a computer scientist and U.S. Navy rear admiral, who found an actual moth trapped in a relay of the Harvard Mark II computer in 1947. The incident was recorded as the "first actual case of bug being found." However, the use of "bug" to describe system glitches or malfunctions predates this incident, with its origins in engineering and hardware systems. Debugging, as a process, emerged alongside the development of more complex computing systems. Over time, as software grew in scale and complexity, debugging evolved into a specialized skill and an integral part of the software development life cycle.

RTL debugging has evolved significantly over the years, adapting to the increasing complexity of digital designs. Initially, designers relied on simple waveform inspection tools and manual tracing to identify functional bugs, often using limited simulation data. As designs grew in size and verification requirements became more rigorous, formal verification techniques, assertions, and coverage-driven methodologies were introduced to improve efficiency and confidence. The emergence of hardware emulation and FPGA prototyping provided faster execution environments, allowing early detection of corner-case bugs. Today, RTL debugging incorporates advanced visualization, root-cause analysis using machine learning, real-time signal tracing, and integrated debug environments that span across simulation, emulation, and post-silicon validation. These developments have transformed RTL debug from a largely manual and reactive process into a more automated, intelligent, and proactive discipline.

2.5 What Is Debugging?

Bugs do not resolve themselves on their own. They require active identification, analysis, and intervention to be fixed. Without the proper debugging process, errors can persist, affecting the functionality and stability of the system. Whether it's a syntax error, runtime issue, or logic flaw, bugs need to be systematically tracked down and addressed to ensure the software or hardware works as intended. Ignoring bugs or hoping they resolve by themselves can lead to more severe problems down the road. Debugging is the process of identifying, analyzing, and fixing bugs or errors within a program. These bugs could range from simple syntax errors to more complex logical or runtime issues that prevent the program from executing them as intended. Debugging is a critical part of software development and occurs after the initial design and functional verification phases. During the design and verification stages, test plans, direct tests, and random tests are used to catch potential bugs. Once an issue is identified, the debugging phase begins with root cause analysis to understand why the error occurred and how to resolve it. Debugging is a skill that is honed over time and through experience. While there are tools and techniques available, such as debuggers, print statements, or logging, effective debugging requires a systematic approach and a deep understanding of the program's logic.

2.6 Why Is Debugging Important?

Debugging is essential because it ensures that the program functions as intended. Without thorough debugging, software may contain critical errors that could lead to unexpected behavior, crashes, or security vulnerabilities. By identifying and fixing bugs early in the development cycle, developers can improve the stability, performance, and reliability of the software. Furthermore, debugging allows developers to understand the deeper workings of their programs and refine their codebase to prevent future issues.

2.7 How Is Debugging Done?

Debugging is an essential part of software development, and it begins with thorough testing. Testing plays a foundational role in detecting errors within the program, helping to uncover runtime and semantic issues that may otherwise remain undetected. While a program may appear to work correctly under normal conditions, comprehensive testing across a variety of scenarios is critical for ensuring that the software functions robustly. Special or boundary cases, such as extreme inputs, empty inputs, or cases where variables are near their limits, are vital in revealing issues that arise under specific conditions. Additionally, testing how the program handles improper or invalid inputs is important, as many bugs emerge from unexpected data or incorrect formats. Once an error is identified, debugging should be approached systematically. It's important to remain patient and analytical; making random changes without understanding the root cause often leads to more issues. A useful technique is to think backwards from the incorrect output by examining it to trace the error back to its source. If the error remains elusive, breaking the program into smaller parts can help isolate the problem area, and simplifying the code by removing recent additions can make it easier to pinpoint where things went wrong. Testing the program with smaller, simpler cases allows for easier identification of errors that might be hidden in more complex scenarios. Additionally, simplifying complex lines of code by breaking them down into smaller, more understandable components can help clarify the logic and make debugging more efficient. By following a methodical approach to debugging, including strategic testing, analyzing output, and simplifying the code, developers can effectively identify and resolve issues in their programs. Debugging spans from RTL design to final verification, involving multiple iterations and collaboration among individuals with varying expertise. This process continues until the code is error-free or meets strict coverage requirements.

2.8 Bugs/Errors Classification

Detecting and correcting errors early in the design and development process is essential to ensure product quality and reduce production costs. Errors can originate from a variety of sources, and they fall into several distinct categories based on their nature, origin, and impact. Broadly, these errors can be grouped into **coding-level errors** (such as syntax, runtime, and semantic issues) and **design-level bugs** (logical, algorithmic, and timing issues), especially relevant in hardware or RTL design contexts (Fig. 2.5).

In the RTL design domain, **critical bugs** are those that directly affect the core functionality of the system, often rendering it non-operational or leading to severe malfunctions. These issues require immediate attention because they can halt progress and prevent successful simulation, synthesis, or implementation. For instance, a **deadlock** in a finite state machine (FSM), caused by incorrect state transitions, or a **race condition** due to improper synchronization of clocks, would be considered critical. Such bugs could result in system crashes or incorrect outputs, making it essential to resolve them before moving forward with further stages of development. Delaying fixes for critical bugs risks system instability and can result in costly project delays.

On the other hand, **non-critical bugs** have a less immediate impact on the system's core functionality and often do not prevent the design from running as intended. These types of bugs are typically related to minor issues that may not significantly alter the overall system behavior but could affect performance or introduce small inefficiencies. For example, a **minor timing violation** that does not lead to functional failure or slight **misalignment of data signals** during simulation could be considered non-critical. These bugs are generally lower priority and can be addressed after more pressing issues are resolved.

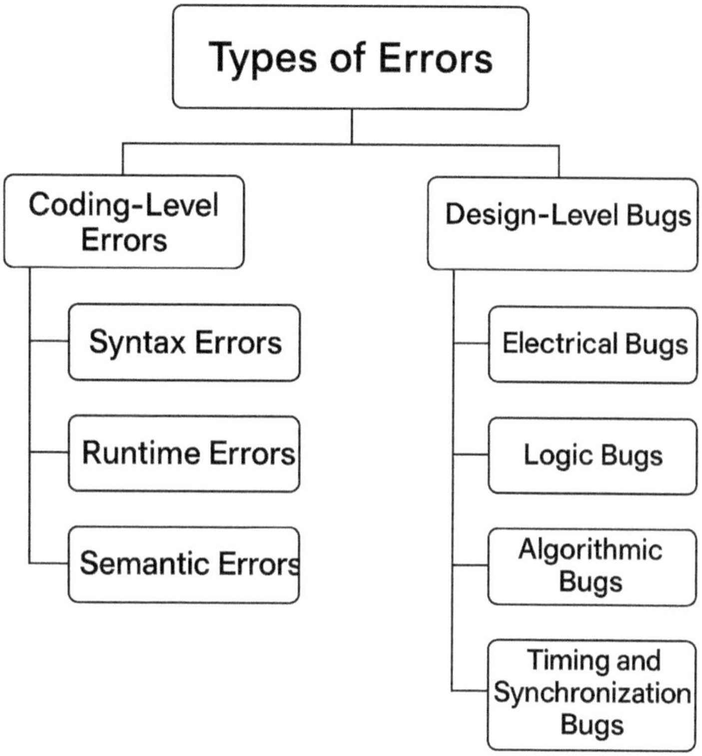

Fig. 2.5 Types of bugs

However, while non-critical bugs do not prevent system operation, if left unresolved, they can accumulate and make the system harder to maintain or optimize in the long run.

2.8.1 Coding-Level Errors

In high-level programming languages like Python, errors typically fall into the following categories:

1. **Syntax Errors**
   These occur when the code violates the grammatical rules of the language. Python flags these errors before execution and provides an error message, though it may not always indicate the exact source of the problem. Missing or mismatched parentheses, brackets, or quotation marks are common causes [14].
2. **Runtime Errors**
   These happen during execution, causing the program to crash. Modeling language provides diagnostic information such as the error type and the line number. However,

Table 2.5 Examples of some runtime errors

Bug type	Classification	Description
Sporadic/ intermittent	Non-deterministic or environmental bug	Triggered by rare, unpredictable conditions; hard to reproduce consistently
Heisenbug (time sensitive)	Timing-related or observability bug	Changes or disappears when observed; often due to timing issues or instrumentation
Race condition	Concurrency bug	Arises from improper handling of shared resources across threads or processes
Buffer overflow	Memory corruption bug	Caused by writing outside the bounds of allocated memory

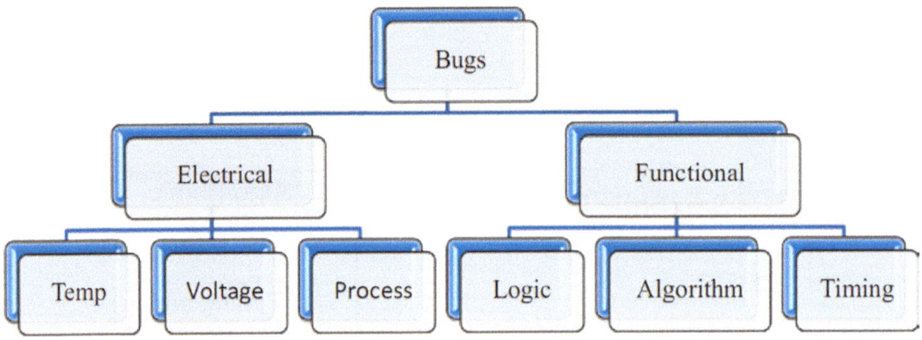

Fig. 2.6 Types of design-level bugs

the root cause may lie earlier in the code. Examples include division by zero or referencing an undefined variable. Examples of some runtime errors are summarized in Table 2.5.

3. **Semantic Errors**

   Semantic errors occur when the program executes without crashing but yield incorrect results due to flawed logic. These errors are the hardest to detect, as they require an understanding of the intended functionality and careful inspection of both the code and its behavior.

2.8.2 Design-Level Bugs

In hardware design and RTL development, functional errors can be more complex and are generally categorized into **electrical bugs** and **Functional bugs** (Fig. 2.6).

1. **Electrical Bugs**

   These are caused by interactions between the design and physical phenomena such as:

   (a) Crosstalk
   (b) Power supply noise

(c) Temperature variations
(d) Process variation
(e) Signal integrity issues

Such bugs are typically discovered during post-silicon validation or detailed physical simulations.

2. **Functional/Logic Bugs**

   These result from incorrect logic in combinational circuits. For instance, using the wrong logic gate or condition can produce incorrect outputs. Logic bugs are typically localized and can often be fixed with targeted code corrections.

3. **Algorithmic Bugs**

   Algorithmic bugs reflect a mismatch between the design's implementation and its specification. These are high-level errors that typically require structural changes to the architecture or control-flow. Fixing algorithmic bugs often involves significant design revision.

4. **Timing and Synchronization Bugs**

   These bugs relate to the temporal behavior of signals. A timing bug might arise if a signal is latched too early or too late, violating expected data transfer timing. Synchronization bugs also include metastability issues across clock domains. Timing bugs may not be easily caught through logic simulation and often require detailed timing analysis or waveform inspection.

2.9 Verilog RTL Debugging: Common Verilog RTL Bug Types

Debugging RTL (Register Transfer Level) designs written in Verilog involves identifying and resolving issues that may not necessarily arise from syntax errors, but from functional misbehavior, timing inconsistencies, or unexpected hardware simulation results. One of the most effective ways to accelerate this process is by categorizing bugs based on their nature, understanding the underlying cause, and recognizing code patterns that frequently lead to failures.

Table 2.6 outlines typical Verilog bug types encountered during RTL design and simulation. For example, width mismatches between buses can silently cause truncation or zero-extension, leading to data corruption that is hard to trace. Logic bugs, such as incorrect use of combinational or sequential statements, may pass simulation yet behave differently in hardware due to race conditions or timing glitches [15].

2.10 Classification of Debugging Methods

2.10.1 Reactive Debugging vs Preemptive Debugging

There are two primary types of debugging approaches: **reactive debugging** and **preemptive debugging**. Both play essential roles in the development process, but they are employed in different contexts and at different stages of the software lifecycle.

Table 2.6 Common Verilog RTL bug types and debugging strategies

Bug type	Description	Original	Bug
Width	Mismatched bit widths causing unintended truncation or extension	input [6:0] data_offset_delay;	input [5:0] data_offset_delay;
Logic	Incorrect logic expression leading to functional or timing bugs	assign rnw = writeback_w;	assign rnw = ~ writeback_w;
Assignment	Improper use of blocking/non-blocking causing race conditions	int_dout_next = 1'b0;	int_dout_next = 1'bx;
Initial	Improper initial value leading to undefined behavior	e_tx_ack <= 1'b0;	e_tx_ack <= 1'b1;
Data	Incorrect handling of signed/unsigned data	wb_sel_o <= #1 cbu_sel_i;	wb_sel_o <= #1 4'b0000;
State	FSM transition or state encoding issues	state <= STATE_INIT;	state <= STATE_START;
Comparison	Misuse of comparison or equality operators	counter == 7	counter != 7
Bitwise	Incorrect use of bitwise operations	PCplusout <= PCplusin;	PCplusout <= PCplusin >> 1;
Condition	Faulty if-else condition or priority encoding	if (pause==0)	if (pause==1)
Signal	Signal declaration issues like naming or type mismatch	assign Out = Switch;	assign Out = Out1;
Arithmetic	Incorrect usage of signed math or underflow/overflow	ndCount <= count + limit;	ndCount <= count – limit;
Value	Wrong parameter or constant assignment	localparam OPCODE = 6'h04;	localparam OPCODE = 6'h05;
Others	Miscellaneous tool or syntax issues	supply1 vddio;	supply0 vddio;

1. **Reactive Debugging**

   Reactive debugging is the process of responding to errors or issues after they have occurred. It involves identifying, isolating, and fixing bugs that arise during program execution or after a failure has been observed. This type of debugging typically happens after the system is deployed or during testing when unexpected behavior or errors are encountered. Reactive debugging is often driven by the need to resolve specific problems that have already been noticed, such as crashes, hangs, or incorrect output. It tends to be a more troubleshooting-oriented approach, focusing on identifying the root causes of these issues, often with the help of debugging tools like simulators, waveform viewers, and log files. Since reactive debugging occurs after the fact, it's often a necessary and time-consuming process, especially for complex systems where the error is not immediately apparent.

2. **Preemptive Debugging**
   Preemptive debugging, on the other hand, aims to prevent errors before they occur. This proactive approach involves identifying potential problems in the design, code, or system architecture before they lead to failures. Preemptive debugging techniques are often applied during the development process, particularly during code writing and early testing phases. This approach includes writing robust test cases, applying static analysis tools, reviewing code for potential pitfalls, and using techniques like assertions or formal verification to ensure that the software behaves as expected under all scenarios. By identifying and addressing issues before they cause failures, preemptive debugging can reduce the time and resources spent on fixing problems later in the development cycle. It can be particularly useful in preventing critical issues in high-stakes environments, where system stability is paramount.

2.10.2 Template-Based vs Heuristic-Based vs AI-Based Debugging

Automatic code debugging is gaining traction as a valuable direction in software engineering. Given a program containing bugs, the goal is to automatically correct the faulty code using appropriate functions, thereby reducing the need for manual debugging and code repair. Traditional methods can generally be grouped into template-based, heuristic-based, constraint-based, and neural network-based techniques [16–23].

1. **Template-Based Debugging**
   Template-based methods rely on expert-defined code patterns to fix bugs but are limited in generalizability. Heuristic-based methods apply predefined rules and may not handle all bug types.
2. **Heuristic-Based Debugging**
   Constraint-based methods solve the repair problem as a constraint satisfaction task, offering precision but often at a high computational cost.
3. **AI-Based Debugging**
   Neural network-based methods require large datasets of labeled code pairs: buggy and corrected examples: for training, which are expensive and time-consuming to collect.

   More recently, large language models (LLMs) have opened new possibilities for automatic code debugging. Trained on vast repositories of open-source code, LLMs are believed to effectively recognize bug patterns and generate correct code fixes. Modern LLM-based approaches often employ retrieval-augmented generation (RAG) and advanced prompt design to enhance their debugging capabilities.

2.11 Understanding the Moments that Matter in RTL Debug

In the ever-growing complexity of System-on-Chip (SoC) design, RTL debugging has evolved into both a science and an art. With massive signal activity, hierarchical modules, and tight integration with software-driven verification, it's easy to get lost in the sea of waveforms and logs. That's why understanding the **"moments that matter"** is critical for efficient and targeted RTL debug [24].

2.11.1 Design Bring-up Events

At the very beginning of every simulation lies a critical sequence of events: reset, clock activation, and configuration. These are often taken for granted but serve as the **root cause** for many downstream bugs.

- **Reset Assertion/De-assertion:**
  Check that every logic block returns to a known state. Trace reset propagation across hierarchy and ensure no flip-flop is left uninitialized.
- **Clock and Power Sequencing:**
  Many IPs rely on gated clocks or staggered power domains. A misconfigured clock mux or power-up sequence can quietly delay functional startup.
- **Software/Host Configuration:**
  If you're simulating a CXL or PCIe device, host-side register programming via MMIO often defines the device's behavior. A wrong bit in a control register might leave a whole IP block disabled.

2.11.2 Protocol Activity: The Basics of Protocol Debugging

Modern SoCs thrive on communication, and that's managed by protocols. These protocols: AXI, TileLink, or CXL: rely on precise **handshakes** and **order guarantees**.

- **Valid/Ready Handshakes:**
  Whether it's an AXI write channel or a CXL.mem flow, observing valid/ready timing reveals throughput bottlenecks, stalls, or dropped packets.
- **Outstanding Transactions:**
  For burst-based protocols, debug becomes exponentially harder if you can't track outstanding reads and writes. Designing debug hooks to track transaction IDs can significantly reduce triage time.

In the context of RTL (Register Transfer Level) protocol debugging, the topology of the interconnect plays a foundational role in how data moves within a system. The topology defines the structural layout and connectivity of components, whether through point-to-point links, shared buses, ring networks, hierarchical interconnects, crossbar switches, or network-on-chip (NoC) routers. Each topology introduces specific behaviors and complexities. For instance, point-to-point interconnects are simpler to debug because they involve fewer parties and clearer data paths. In contrast, shared buses or NoCs can introduce contention, congestion, or packet misrouting. During debugging, engineers typically trace signal activity across modules to determine whether data packets or transactions are being correctly routed. If a master initiates a transaction but the slave does not respond, the issue may be within the routing logic, arbitration, or address decoding in the interconnect. Visual tools like waveforms, transaction-level logs, and router state inspections help verify that signals are reaching their destinations as intended and that no packets are being dropped or looped indefinitely within the interconnect fabric.

Arbitration is another critical element in RTL protocols that governs which master gains access to a shared communication medium at any given time. This is especially important in systems where multiple initiators contend for a common resource, such as a shared memory or peripheral bus. Arbitration schemes can vary, ranging from static priority systems where certain masters always have precedence, to dynamic schemes like round-robin, token-passing, TDMA (Time Division Multiple Access), or lottery-based methods. During RTL debugging, it is important to verify that the arbitration logic is behaving correctly and fairly, particularly under high traffic conditions. Misbehavior can lead to starvation, where low-priority masters never get access, or to deadlocks if arbitration decisions are not properly resolved. Debugging involves checking the request and grant signals in the design, observing whether the correct master is selected based on the arbitration policy, and confirming that no request is ignored or indefinitely postponed. This often requires simulating scenarios with overlapping and simultaneous requests and inspecting the internal state of arbiter finite state machines (FSMs).

The bus width, encompassing the data, address, and control buses, directly influences the quantity and granularity of information that can be transferred in a single cycle. In RTL design, mismatches in bus width between masters and slaves can cause serious issues. For instance, a 64-bit master communicating with a 32-bit slave may inadvertently cause truncation, misalignment, or even invalid memory accesses if proper conversion or adaptation logic is not implemented. Debugging such issues involves careful inspection of signal assignments and parameter configurations to ensure that bus widths are harmonized across modules. Furthermore, assertions can be embedded into the design to check for width mismatches or to flag the appearance of undefined or floating signals, such as x or z values. Waveform analysis also provides insight into whether full data payloads are being correctly transmitted and received, and whether address decoding logic is interpreting the signal widths accurately.

The mechanism of data transfer—whether pipelined, non-pipelined, or burst—defines how efficiently and reliably transactions are executed in the protocol. Pipelined transfers allow overlapping of transactions, where new commands can be issued before previous ones complete, thereby improving throughput. Non-pipelined systems, by contrast, require that one transaction complete before the next begins, which simplifies control logic but reduces performance. Burst transfers involve sending a series of data beats in a sequence, often with a single address phase followed by multiple data phases. Each of these modes introduces its own set of debugging challenges. In pipelined systems, hazards can arise if data dependencies are not properly managed, or if control signals such as VALID and READY are not correctly synchronized. In burst transfers, debugging focuses on ensuring that the burst length, alignment, and data consistency are preserved. Simulation waveforms help detect if the burst is interrupted prematurely, if beat ordering is corrupted, or if the recipient module does not properly latch all data words.

Timing is a core consideration in digital systems and becomes particularly complex when multiple clock domains or asynchronous elements are involved. In RTL designs, timing can be either synchronous—where all components share a common clock—or asynchronous—where different modules operate on unrelated clocks. Asynchronous systems introduce the risk of metastability, a condition where signal transitions violate setup or hold timing constraints, resulting in indeterminate logic states. These issues are notoriously difficult to catch through simulation alone, which is why formal CDC (Clock Domain Crossing) analysis tools are commonly employed. Debugging timing issues involves verifying that synchronizer, such as double flip-flops or FIFO interfaces, are present and correctly implemented at clock domain boundaries. Additionally, asynchronous resets must be deasserted in a controlled and synchronous manner to avoid glitches or stuck-at conditions. Engineers analyze timing paths in waveform viewers, check reset synchronization logic, and often rely on linting tools to flag potential violations early in the design cycle.

Transmission control defines how transactions are initiated, acknowledged, and completed within a protocol. This is typically handled through handshaking mechanisms such as the VALID and READY signals in AXI, or REQ and ACK in simpler protocols. These signals ensure that data is only transferred when both the sender and receiver are ready, preventing data loss or overwrites. Other protocols may rely on preambles—special headers that precede the data to define its type, size, or destination. Debugging transmission control requires validating that handshaking protocols are respected at all times. For instance, a master must assert VALID only when it has data to send and must wait until the slave asserts READY before considering the data transferred. If either side fails to adhere to this sequence, transactions may be dropped or erroneously accepted. Waveform analysis is crucial here, allowing engineers to visually align the control signals and ensure that the proper sequencing is followed. Protocol checkers or embedded assertions can also automatically detect violations in the handshaking protocol.

Finally, the type of transmission—be it point-to-point, multicast, or broadcast—determines the reach and nature of the communication. Point-to-point links are straightforward, involving one sender and one receiver. However, multicast and broadcast introduce more complex requirements, such as simultaneous delivery to multiple receivers, signal replication, and acknowledgment aggregation. RTL debugging in these scenarios involves ensuring that all intended recipients receive the correct data, that address decoding supports group targets, and that no unintended modules respond to a message. Problems often arise when one receiver drops data due to a readiness issue or when signal fan-out causes delays or race conditions. Engineers must inspect each receiver's behavior, use coverage analysis to determine message reach, and verify that each transmission path satisfies the protocol's delivery guarantees. Assertions and simulation logs are particularly useful for verifying that broadcast messages are propagated and received consistently across the system.

2.11.3 Control and Status Signal Transitions

This category involves signals that govern **when** and **how** a unit should behave—enable bits, done pulses, start triggers, and interrupts.

- **Start-of-Operation to Done:**
  Bookmark these events to measure performance or validate behavior. Did a matrix multiplication complete in 100 cycles? Why did it take 400?
- **Interrupt Generation and Servicing:**
  A mismatch in interrupting timing or polarity can derail software-hardware interaction. Trace back from interrupted flags to root events like FIFO thresholds or error conditions.

2.11.4 FSM Behavior: The Heartbeat of Control Logic

Finite State Machines (FSMs) often encapsulate core control logic. An unexpected transition—or worse, a stuck state—can silently stall a block.

- **Illegal States:**
  Incorporate assertions that fire when FSMs enter undefined states. These are goldmines for uncovering synthesis or reset bugs.
- **Livelock/Deadlock Detection:**
  If you see a system "running" but not "progressing," suspect circular wait states in FSMs or blocking conditions on shared resources.

2.11.5 Checker Mismatches and Testbench Anomalies

Many testbenches include scoreboards, reference models, or protocol checkers. A **checker mismatch** is more than an error: it's a clue to the exact moment something went off-script.

- **First Mismatch Principle:**
  Always trace the *first* mismatch rather than cascading failures. The earlier bug is often the root cause.

- **Assertion Failures:**
  Time-stamped and localized, assertions give pinpoint insight into violated expectations. A good assertion is a future time machine for your debug efforts.

2.11.6 Power and Performance Monitors

Modern chips are power-sensitive and performance-hungry. Watching signals related to **clock gating**, **power domains**, and **latency counters** tells you if you're meeting spec—or just wasting cycles.

- **Gating Transitions:**
  Was the clock gated when the block should've been idle? Did power gating disable the RAM before its final access?
- **Latency Peaks:**
  Use performance counters to log latencies of critical paths. Match these peaks to activity windows in waveform to spot contention or arbitration issues.

2.11.7 Software-to-Hardware Interactions

For software-driven verification, SW actions translate into observable HW behavior. Understanding this **translation bridge** is crucial:

- **Trigger Writes:**
  When software writes a "start" bit or a command queue entry, what hardware events should follow? Is there a timeout?
- **Status Reads:**
  Software often polls for completion. If status flags aren't updated, check whether the backend FSM stalled or if a writeback didn't happen.

2.12 RTL Design Languages

Register Transfer Level (RTL) design is the foundation of digital circuit development, enabling precise behavioral and structural descriptions of hardware systems. RTL design languages are used to model, simulate, verify, and ultimately synthesize logic for implementation in ASICs or FPGAs. The most common RTL languages include Verilog, VHDL, and SystemVerilog, each serving different strengths and historical preferences in the semiconductor industry.

2.12.1 Verilog

Verilog is one of the earliest and most widely used hardware description languages (HDLs). It uses a C-like syntax, making it relatively easy for software engineers to learn. Verilog supports structural, behavioral, and gate-level modeling [25].

Strengths:

- Simple syntax for combinational and sequential logic.
- Strong industry tool support.
- Widespread usage in IP and SoC design.

Use Cases:

- ASIC and FPGA RTL design.
- Testbench creation.
- Synthesis and gate-level simulation.

2.12.2 VHDL

VHDL (VHSIC Hardware Description Language) was developed under a U.S. Department of Defense program. It is strongly typed and verbose compared to Verilog, promoting strict design discipline and reliability.

Strengths:

- Rigid syntax reduces ambiguity.
- Strong typing helps catch errors at compile time.
- Suitable for safety-critical and aerospace applications.

Use Cases:

- RTL design in avionics and defense.
- Formal verification environments.
- Documentation-rich IPs.

2.12.3 SystemVerilog

SystemVerilog extends Verilog with powerful verification constructs and advanced RTL features. It is now the de facto standard for both design and verification.

Strengths:

- Unified language for RTL and verification.
- Object-oriented features and constrained random generation.
- Assertions, interfaces, and coverage support.

Use Cases:

- SoC design and verification.
- UVM-based testbenches.
- Complex interconnect modeling (e.g., AXI, CXL).

2.12.4 Emerging and Domain-Specific Languages

Recent years have seen the rise of domain-specific and high-level synthesis (HLS) languages, aiming to abstract hardware design and increase productivity.

Examples:

- Chisel: A Scala-based HDL that enables parameterized and reusable hardware generation.
- Bluespec: Based on term-rewriting rules, used for rapid prototyping.
- C/C++ HLS: Converts C-like descriptions to RTL using synthesis tools (e.g., Vivado HLS).

Use Cases:

- Academic research and teaching.
- AI accelerators and custom data paths.
- Fast iteration and prototyping.

2.12.5 Language Interoperability and Mixed-Language Simulation

As RTL designs become more complex, mixed-language support is critical. Most modern simulators allow integration of Verilog, VHDL, and SystemVerilog in a single testbench environment.

Best Practices:

- Use wrappers and bindings to interface between languages.
- Maintain consistent naming and hierarchy conventions.
- Leverage simulator features like foreign function interface.

2.13 Applying Good vs Bad Debugging to RTL Design

What Bad RTL Debugging is:

- **Add console logs everywhere**:
  In RTL, this translates to flooding waveforms or testbenches with $display or VCD outputs, making it hard to isolate root causes.
- **Debug on latest git commit:**
  RTL regressions are often introduced subtly. Debugging only the tip-of-branch RTL can mask when/where bugs were introduced. You miss the opportunity to use tools like git bisect on behavioral or structural netlist commits.
- **Debug alone for hours:**
  RTL bugs can be deeply layered—involving protocol, timing, reset syncs, etc. Isolating yourself can cause tunnel vision, especially without architectural insights or domain knowledge.
- **Make assumptions about the bug:**
  A typical example: assuming the issue is timing-related without checking if the testbench is initializing the DUT correctly.
- **Make random changes to fix the bug:**
  Tweaking FSM states or flip-flop resets without understanding signal dependencies or CDC paths often introduces new bugs.
- **Skip reproducibility steps:**
  If you don't regenerate waveforms from a minimal failing pattern or don't constrain the testbench for deterministic behavior, you can't isolate what's really broken.

What Good RTL Debugging is:

- **Use a debugger to save time:**
  This includes using waveform viewers like DVE, Verdi, or GTKWave with signal tracing and time markers, not just raw logs.
- **Use git bisect to find bug faster:**
  Bisect between RTL commits or even testbench commits to find where the signal behavior diverges. Combine with waveform diff tools.
- **Ask a rubber duck about the bug:**
  Literally explaining the RTL block diagram, signal interactions, or FSM transitions to a colleague or even to yourself often surfaces flawed assumptions.
- **Check logs to find the bug:**
  This means not only simulation logs (xprop, x-check, $fatal, etc.) but also synthesis reports, lint violations, and CDC/DFT checks.
- **Systematically fix the bug:**
  Create a minimal test that reproduces the failure. Apply focused changes like adding one reset signal path or gating one clock, then observe the result.

- **Reproduce the failure to ensure bug is fixed:**
  After applying the fix, rerun both the minimal and full testbench environments. Checkwaveforms, monitor coverage, and rerun lint/static checks to ensure no collateral damage.

2.14 A Six Sigma Methodology for RTL Debugging

In today's rapidly evolving semiconductor industry, RTL design, verification, and debugging remain critical bottlenecks in time-to-market. As SoCs grow more complex, efficient functional debug and bring-up methodologies become essential. Traditional approaches often involve redundant effort, inconsistent debug flows, and extensive manual analysis. These inefficiencies directly impact productivity and product quality [26–29].

The Six Sigma methodology, originally developed by Motorola in 1986, offers a structured data-driven approach to reduce process variation and improve quality. Using its five-phased DMAIC framework—**Define, Measure, Analyze, Improve, and Control**—Six Sigma can identify root causes of design inefficiencies and standardize debugging flows to ensure consistent, high-quality deliverables across verification and bring-up teams (Table 2.7).

2.14.1 DMAIC Implementation

1. **Define Phase**
   This phase identifies candidate processes for improvement—such as waveform analysis flows, testbench initialization, signal observability, or SDABE tool integration. Inputs to this phase include debug delay metrics, test coverage gaps, or feedback from post-silicon teams. Prioritization is based on ROI, customer-impact severity, and scalability.

Table 2.7 DMAIC application in RTL debug

Phase	Purpose	Example tools used
Define	Identify debug inefficiencies and goals	Process Maps, Pareto Charts
Measure	Collect metrics on debug/simulation flow	Check Sheets, Histograms
Analyze	Find root causes for bottlenecks/errors	Control Charts, Fishbone Diagrams
Improve	Implement reusable and automated solutions	Scatter Plots, Cost-Benefit Tools
Control	Validate and institutionalize improvements	Flow Charts, Control Dashboards

2. **Measure Phase**

    Accurate data collection is essential. This includes measuring:
    (a) RTL simulation time per testcase.
    (b) Bug root cause isolation time.
    (c) Redundancy in wave dump signal lists.
    (d) Efficiency of automated bring-up scripts.

Using tools like check sheets, histograms, and Pareto charts, teams can visualize where bottlenecks exist in RTL debug or SDABE flows.

3. **Analyze Phase**

    At this stage, debug logs, waveform reviews, and protocol trace alignment data are analyzed to identify root causes of recurring issues:
    (a) Unobservability of internal signals.
    (b) Late detection of protocol violations.
    (c) Inconsistent simulation environment setups.

Control charts and fishbone diagrams help trace these issues back to structural weaknesses.

4. **Improve Phase**

    Once inefficiencies are identified, improvements are proposed and implemented:
    (a) Creating generic monitor/checker modules.
    (b) Automating environment bring-up (reset, config, stimulus).
    (c) Consolidating signal trace lists.
    (d) Enabling waveform bookmarks for common issues.

Scatter plots can help evaluate correlation between debug time and testcase complexity.

5. **Control Phase**

    After deploying improvements, it's crucial to monitor ongoing projects using control charts and periodic audits to ensure improvements are retained. A version-controlled infrastructure, debug dashboard, and reusable issue templates support long-term maintainability and scale.

References

1. Smith, J., & Jones, M. (2020). Solving-oriented and domain-oriented knowledge structures: Their application to debugging problem solving activity. *Cognitive Processes in Debugging, 12*(2), 35–48.
2. Gansel, C. (2021). *The art of designing embedded systems*. Wiley.
3. Kernighan, B. W. (1999). *The practice of programming*. Addison-Wesley.
4. Retrieved from https://opentext.wsu.edu/psych105/chapter/7-4-problem-solving
5. Sannemo, J. (2019). *Principles of algorithmic problem solving*. KTH Royal Institute of Technology.
6. Kerschke, P., Hoos, H. H., Neumann, F., & Trautmann, H. (2018). Automated algorithm selection: Survey and perspectives. *arXiv:1811.11597v1*.
7. Gupta, S. K. (2023). IX. Algorithmic paradigms. In *Data structures and algorithms*. Springer.
8. O'Reilly, J. K. (2015). Cognitive and computational complexity: Considerations from the perspective of problem-solving. In *Springer handbook of computational intelligence* (pp. 123–145). Springer.

9. Kerschke, P., et al. (2014). A survey of algorithmic skeleton frameworks: High-level structured parallel programming. *Software, Practice & Experience, 44*(6), 769–802.
10. Kumar. (1992). Algorithms for constraint-satisfaction problems: A survey. *AI Magazine, 13*(1), 32–44.
11. Sedgewick, R., & Wayne, K. (2011). *Algorithms* (4th ed.). Addison-Wesley/Pearson.
12. Zámecniková, V., & Hromkovic, J. (2006). *Design and analysis of randomized algorithms: Introduction to design paradigms*. Springer.
13. Gendreau, M., & Potvin, J.-Y. (Eds.). (2010). *Handbook of metaheuristics* (2nd ed.). Springer.
14. Tsai, Y., Liu, M., & Ren, H. (2024). Rtlfixer: Automatically fixing RTL syntax errors with large language model. In *Proceedings of the 61st ACM/IEEE Design Automation Conference, DAC*. ACM.
15. Mohamed, K. S. (2024). HW/SW co-verification and co-debugging. In *Heterogeneous SoC design and verification: HW/SW co-exploration, co-design, co-verification and co-debugging, synthesis lectures on digital circuits & systems* (pp. 133–151). Springer.
16. Wen, M., Chen, J., Wu, R., Hao, D., et al. (2018). Context-aware patch generation for better automated program repair. In *Proceedings of the 40th International Conference on Software Engineering*. ACM.
17. Xiong, Y., Wang, J., Yan, R., Zhang, J., Han, S., Huang, G., & Zhang, L. (2017). Precise condition synthesis for program repair. In *2017 IEEE/ACM 39th International Conference on Software Engineering (ICSE)* (pp. 416–426). IEEE.
18. Xuan, J., Martinez, M., Demarco, F., Clement, M., Marcote, S. L., Durieux, T., Le Berre, D., & Monperrus, M. (2016). Nopol: Automatic repair of conditional statement bugs in java programs. *IEEE Transactions on Software Engineering, 43*(1), 34–55.
19. Yuan, Y., & Banzhaf, W. (2018). Arja: Automated repair of java programs via multi-objective genetic programming. *IEEE Transactions on Software Engineering, 46*(10), 1040–1067.
20. Zhang, X., Zhai, J., Ma, S., & Shen, C. (2021). Autotrainer: An automatic dnn training problem detection and repair system. In *2021 IEEE/ACM 43rd International Conference on Software Engineering (ICSE)* (pp. 359–371). IEEE.
21. Chen, Z., Kommrusch, S., & Monperrus, M. (2022). Neural transfer learning for repairing security vulnerabilities in c code. *IEEE Transactions on Software Engineering, 49*(1), 147–165.
22. Jiang, J., Xiong, Y., Zhang, H., Gao, Q., & Chen, X. (2018). Shaping program repair space with existing patches and similar code. In *Proceedings of the 27th ACM SIGSOFT International Symposium on Software Testing and Analysis*. ACM.
23. Tsai, Y. D., Liu, M., & Ren, H. (2023). RTLFixer: Automatically fixing RTL syntax errors with large language models. *arXiv preprint arXiv:2311.16543*.
24. Chang, K., Markov, I. L., & Bertacco, V. (2009). *Functional design errors in digital circuits*. Springer.
25. Mohamed, K. S. (2016). Verilog for implementation and verification. In *IP cores design from specifications to production. Analog circuits and signal processing*. Springer.
26. Whittaker, J. A., & Voas, J. M. (2002). 50 years of software: Key principles for quality. *IT Professional, 4*, 28–35.
27. Klefsjö, B., & Bergman, B. (2004). *Quality from customer needs to customer satisfaction*. Studentlitteratur.
28. Magnusson, K., Kroslid, D., & Bergman, B. (2000). *Six sigma - The pragmatic approach*. Studentlitteratur.
29. HCi Library. (2008). Cause and effect diagrams. Retrieved from http://hci.com.au/hcisite5/library/materials/Cause%20and%20effect%20diagrams.htm

3 Conventional Methods for RTL Debugging "Rule-Based Techniques"

3.1 Before Debugging: Laying the Groundwork for Efficient RTL Development

Effective debugging begins long before the first error is encountered. Proactive design practices, such as careful planning, modular development, and the use of expressive constructs like enumerated states, help minimize the likelihood of bugs and simplify issue identification when they arise. Rather than diving directly into debugging, engineers should first consider whether they're observing a root problem or just a symptom, as misdiagnosing the cause can lead to wasted effort and recurring failures. Distinguishing symptoms from true faults is essential for applying meaningful, lasting fixes. Additionally, maintaining a structured bug tracking system allows teams to efficiently manage both familiar and novel issues. Known bugs can often be resolved quickly using past experience and logs, while new bugs require investigative techniques such as signal tracing, assertion checking, and waveform analysis. Overall, thoughtful coding practices, critical diagnostic reasoning, and systematic issue tracking form the foundation for successful RTL debugging.

3.1.1 Prevention Is the Best Medicine

To minimize the likelihood of errors, it's important to write code carefully and correctly from the start. Additionally, the code should be structured in a way that makes it easier to identify and fix any issues that do occur. For instance, using clear data representations such as defining states with enum integer {step1 = 0, step2 = 1, step3 = 2} state; makes it easier to interpret waveforms during simulation. When a bug is discovered, it's important not to jump straight into debugging. First, ask yourself key questions: Could this issue exist

K. Mohammed, *The Art of RTL Debugging*, Synthesis Lectures on Digital Circuits & Systems, https://doi.org/10.1007/978-3-032-15880-2_3

elsewhere? What other bugs might be masked by this one? How can I prevent similar bugs in the future? Once you've considered these, begin debugging through waveforms, logging, or tracing. To further enhance robustness, incorporate mechanisms like a built-in watchdog timer class to automatically detect and handle testcase hangs. Before you start coding, it's important to plan ahead and think carefully about the overall structure of your program. Developing and testing small parts of the program helps catch issues early, as you can often run your code after writing just a few lines, even if it doesn't seem like a natural stopping point. Maintaining a consistent and clean coding style also reduces the chances of errors and makes any mistakes easier to identify and fix.

3.1.2 Are You Looking at the Problem or at a Symptom?

When debugging, it's essential to distinguish between looking at the problem itself and focusing on a symptom. A symptom is often an observable effect or outcome caused by an underlying issue, but it may not directly point to the root cause. For example, a system crash might be a symptom, but it doesn't necessarily reveal the reason behind it; there could be multiple potential causes, such as memory leaks, race conditions, or incorrect logic.

On the other hand, the problem refers to the actual root cause, which is the underlying issue that leads to the symptoms. It's crucial to address the root cause rather than merely fixing the symptoms, as doing so ensures that the solution is comprehensive and prevents the issue from recurring. If you focus on symptoms, you might temporarily resolve the issue, but the underlying cause will remain, leading to future failures.

3.1.3 Efficient Strategies for Handling New and Known Bugs in RTL Debugging: Bug Tracking System

When debugging RTL designs, it is essential to distinguish between new bugs and those that have been previously encountered. Known bugs often follow familiar failure patterns or stem from well-understood design or integration issues, allowing engineers to apply prior solutions or reference past debug logs and fixes. These can usually be addressed more quickly by leveraging documentation or simulation test cases from earlier projects. On the other hand, new bugs require a more investigative approach; starting from symptom analysis, narrowing down the signal or module at fault, and systematically applying waveform inspection, assertions, or checkers to isolate the root cause. Maintaining a well-organized **bug tracking system** and a history of debug cases can greatly accelerate the resolution of both new and old bugs, while also promoting knowledge sharing across the team.

3.2 Debugging Levels: SW Debugging vs HW Debugging

To effectively debug a system, it is crucial to have both good controllability and observability. This involves probing the system with instruments to stimulate and observe the hardware behavior, as well as instrumenting the software to cause and monitor its actions. It is essential to inject stimulus and carefully observe the results, while keeping detailed notes of any observations made. Hardware instruments such as oscilloscopes, and signal generators can be used to probe and measure various signals in the system. On the software side, tools like printf statements, and monitoring a timer can provide valuable insights into the system's behavior and performance.

Debugging techniques vary significantly between different levels of hardware and software, each with its own set of tools and methodologies. The table below (Table 3.1) outlines how various debugging techniques are applied across different levels, from software (SW) to post-silicon validation. Each level in the development process presents unique challenges, and the debugging techniques must be adapted accordingly. These techniques will be discussed in the following subsections [1].

3.2.1 Software Debugging

Software debugging typically benefits from higher observability and more interactive control compared to hardware. Debugging can be performed using software debuggers, IDEs, and logging mechanisms, and it is often iterative and immediate.

Table 3.1 Debugging techniques vs different levels

	SW	HW					
Debugging technique	SW	TLM	RTL	Gate-level	SoC-level	Physical-level	Post-silicon-level
Waveform analysis	✗	✗	✓	✓	✓	✓	✓
Assertions	✓	✓	✓	✓	✓	✓	✓
Printf-style debugging	✓	✓	✓	✗	✓	✗	✓
Code-level tracing	✓	✓	✓	✗	✓	✗	✓
FSM debugging	✗	✗	✓	✓	✓	✗	✓
Clock/reset issues	✗	✗	✓	✓	✓	✓	✓
Netlist probing	✗	✗	✗	✓	✗	✓	✓
Timing violation debugging	✗	✗	✗	✓	✓	✓	✓
Gating and glitch analysis	✗	✗	✗	✓	✓	✓	✓
Protocol-aware debugging	✗	✓	✓	✓	✓	✓	✗
Protocol analyzer device	✗	✗	✗	✗	✓	✗	✓

Key software debugging techniques include:

- **Printf-style debugging**: Inserting printf or log statements at key points in the code to display variable states, control flow, or event timing.
- **Code-level tracing**: Using tools or instrumentation to capture the sequence of function calls and operations during runtime.
- **Breakpoints and watchpoints**: Pausing execution at a specific point in a debugger to inspect variable values and processor state.
- **Software timers and counters**: Tracking how long specific operations take, which is useful for identifying performance bottlenecks.
- **LEDs and GPIO toggling**: For embedded software, toggling output pins or LEDs helps indicate execution progress or state transitions when no serial output is available.

These techniques provide high control over software execution and are especially effective in early-stage software development, firmware debugging, and driver bring-up.

3.2.2 Hardware Debugging

Hardware debugging is inherently more complex due to lower visibility into the system's internal state and limited ability to instrument once the design is synthesized or fabricated.

Essential hardware debugging tools and methods include:

- **Waveform viewers**: Visualizing signal transitions over time in RTL or gate-level simulations.
- **Assertions (SVA)**: Embedding temporal properties in RTL code to detect violations of expected behavior.
- **Oscilloscopes and logic analyzers**: Capturing physical signal waveforms on boards to analyze timing, glitches, or protocol activity.
- **Netlist probing and signal tracing**: Inserting probes into synthesized netlists or using FPGA logic analyzers (e.g., SignalTap, Chipscope) to capture internal signals.
- **Clock/reset issue detection**: Verifying the correctness of clock domains, synchronization circuits, and reset sequencing.
- **Protocol-aware debuggers**: Tools that understand and decode specific bus protocols (e.g., AXI, PCIe) to help identify transaction-level issues.
- **Emulation and FPGA prototyping tools**: Platforms that allow running the design at or near real-time to catch bugs not seen in simulation [2].

3.2.3 SW/HW Interface Debugging: UART, GDB and JTAG

Debugging across the software and hardware boundary is often one of the most complex and critical aspects of embedded system validation. The interface between software and hardware, where firmware interacts with peripherals, registers, and memory-mapped I/O, can expose synchronization issues, initialization errors, and protocol violations. Tools like GDB (GNU Debugger) are widely used in software to step through code, inspect variables, and set breakpoints at runtime. When combined with JTAG (Joint Test Action Group) interfaces, engineers can gain low-level control of hardware, halt processors, and access registers and memory directly. JTAG serves as a physical interface that allows non-intrusive access to the target system for debugging and boundary scan testing. Together, GDB and JTAG provide a powerful mechanism for cross-domain visibility, enabling step-by-step code execution alongside hardware signal monitoring. This tight integration is essential for resolving elusive bugs where the root cause might lie in the timing, register configuration, or handshaking between the processor and the hardware IPs.

GDB is a widely used debugger developed as part of the GNU project and supports multiple programming languages, including C, C++, and Java. It is a familiar tool for most software developers and plays a crucial role in identifying and resolving software malfunctions. GDB retrieves detailed runtime information, allowing developers to pinpoint the root causes of bugs. One of its significant features is the Remote Serial Protocol (RSP), which enables GDB to connect to remote targets that implement the server side of RSP—this is especially useful in embedded systems, where GDB cannot run directly on the target device. To use GDB for debugging a C program, the code must first be compiled with debugging symbols by including the -g flag in the compilation command. After compilation, the program is loaded into GDB via the terminal. Developers can then set breakpoints at specific lines or functions where they suspect issues. Running the program through GDB will cause it to pause at those breakpoints, providing real-time information such as variable values and the call stack. Developers can continue debugging by stepping through the code using commands like next to move to the next line or print to display the value of variables. This interactive and informative approach makes GDB an essential tool for diagnosing and fixing software bugs efficiently.

JTAG, or Joint Test Action Group, is a widely adopted standard for testing and debugging digital circuits, particularly in complex digital systems. First introduced in the 1980s, it has since become a common feature in many integrated digital devices. At the core of JTAG is the boundary scan chain, a series of logic connections that link all the integrated circuits on a board. This chain enables the sequential input and output of signals through each device, allowing thorough inspection and control. The JTAG interface typically comprises four primary signals: TCK (Test Clock) for synchronizing data transfers, TMS (Test Mode Select) for controlling the state of the test logic, TDI (Test Data In) for feeding test data into the system, and TDO (Test Data Out) for capturing the output data. A major advantage of JTAG is its capability to access internal registers of devices, even when these are not physically reachable. This feature makes it an essential tool for developers, enabling in-depth testing, debugging, and validation of hardware designs at the register level.

UART as a **Debugging Interface** is a widely used technique in SoC and embedded system development, especially under the category of Low-Level Software and Bring-Up Debugging. It provides a lightweight, reliable, and non-intrusive method to monitor system behavior, print diagnostic messages, and interact with the design during runtime. Engineers often integrate UART early in the boot flow to log firmware status, report error conditions, or validate hardware initialization sequences. It's also used to implement simple command-line shells that allow direct register reads/writes or memory dumps, helping debug subsystems without full JTAG or scan support. Additionally, some designs embed UART transmitters inside RTL blocks to trace FSM transitions or trigger-specific debug events, especially valuable in emulation or FPGA prototypes where traditional waveform access is limited. Due to its minimal hardware requirements and ease of integration, UART remains a go-to solution for foundational debug infrastructure during SoC validation and bring-up [3].

Why These Protocols Matter for RTL Debugging?

- **Observation**: Directly read internal flip-flops, FIFOs, bus transactions, and memory interfaces during dynamic operation; without invasive modifications.
- **Control**: Halt or single-step design execution, set breakpoints, or inject stimuli right into RTL registers or logic.
- **Trace**: Record signal activity over time to analyze behavior through layers of the design (especially with high-speed trace protocols).
- **Portability and Scalability**: Use universal connectors/standards across multiple boards, FPGAs, or ASIC prototypes.
- **Tool Interoperability**: Supported across vendors (Xilinx, Intel, ARM, MIPS, etc.) and debug tools, facilitating smooth integration into EDA environments.

3.3 RTL-Based Debugging Techniques: High-Level

Figure 3.1 summarizes the most famous debugging techniques at RTL-Level.

3.3.1 Review the Basics

When debugging digital systems, several foundational aspects must be thoroughly reviewed to ensure a stable and observable environment. First and foremost is the clock (Clk); any irregularities, glitches, or incorrect frequencies can cascade into unpredictable system behavior. Tools such as oscilloscopes or waveform viewers are essential for verifying clean clock signals. Similarly, a properly sequenced and synchronized reset (RST) is critical. Many bugs stem from incomplete or asynchronous reset behavior, which may

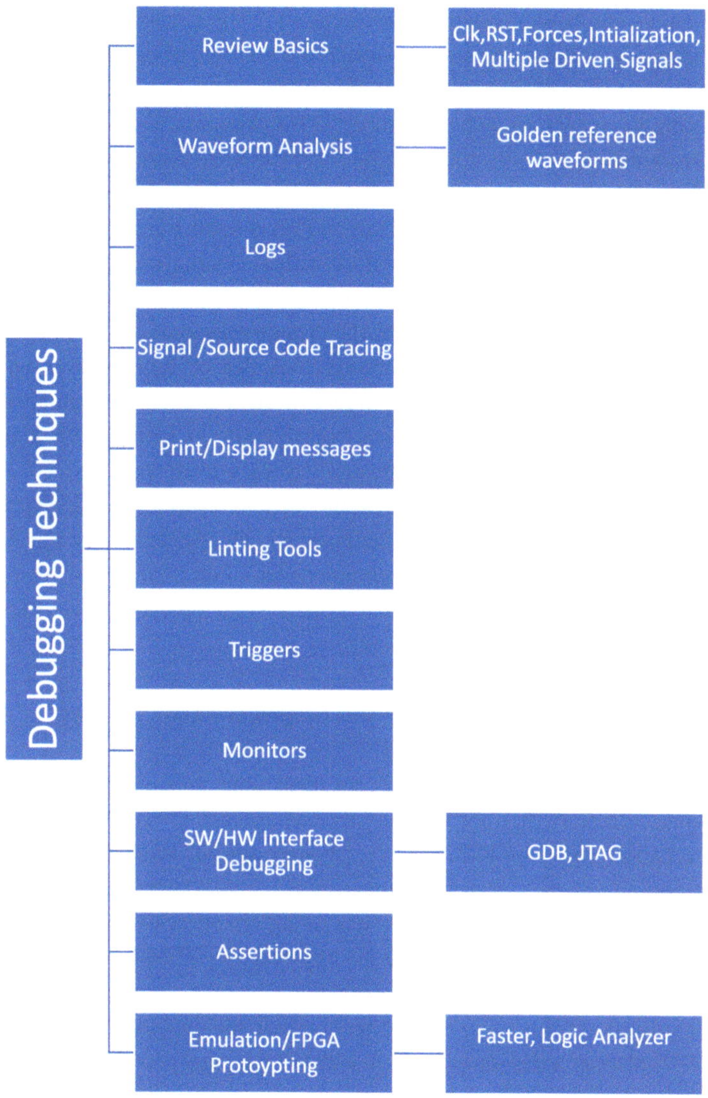

Fig. 3.1 Debugging techniques

leave modules in undefined states. Another key aspect is the use of initialization routines. Registers, memories, and outputs should be explicitly initialized to known values to prevent startup anomalies. In simulation or emulation environments, forcing signals (that is, manually setting signal values) can help isolate and reproduce specific conditions for debugging. However, care must be taken as they may bypass actual design logic. Also, attention must be given to multiple driven signals, which often indicate contention on internal or external buses. Such issues can lead to x-propagation in simulation or functional failures in silicon. Systematic verification of these basics—Clk, RST, initialization, signal integrity, and drive conflicts—forms the first line of defense in identifying and resolving many common design and integration bugs.

Using a sampling clock as a check during RTL debugging is a common and effective technique for identifying timing-related issues, protocol violations, or unexpected behavior in digital designs. A sampling clock is essentially a separate clock signal, often slower than the main design clock, used to observe and record the state of important signals at regular intervals. This helps engineers verify whether specific events occur at the correct times and whether signals like data, valid, or ready behave as expected. It is especially helpful for confirming correct operation of handshake protocols or detecting glitches and missed transitions. In testbenches, a sampling clock can be implemented with a simple toggling process and used to drive logging or assertions. For example, engineers might use it to continuously print out the status of a state machine or the values of critical data paths, making it easier to identify where the design diverges from expected behavior. When debugging clock domain crossings or asynchronous interfaces, the sampling clock can help verify that signals are stable and synchronized. This technique works well in both simulation and hardware emulation environments and complements waveform viewers and assertions by providing a readable trace of the system's behavior over time.

3.3.2 Waveform Debugging: Manual Inspection for Tracing

At the software level, waveform analysis is not applicable, as it requires observing electrical signals and behaviors that are not directly represented in software. However, at the hardware levels; ranging from RTL to physical-level design; waveform analysis becomes crucial. It allows engineers to observe signal timings, check signal integrity, and diagnose timing violations or glitches. Post-silicon debugging also benefits from waveform analysis, as real-world signals are measured on the physical chip after fabrication.

RTL code specifies how signals in wires and registers are computed at each clock cycle. To debug a hardware design, engineers either simulate the design in software or deploy it on an FPGA, capturing a subset of signals with a logic analyzer. Both approaches yield waveform traces that record the values of monitored signals on every clock cycle. These traces can be visualized using waveform viewers, which allow designers to navigate through time by zooming and panning. However, because waveforms reflect RTL-level behavior, they can be difficult to relate to bugs originating from higher-level source languages compiled down to RTL. Temporal localization of bugs is challenging, as time is measured in clock cycles rather than in terms of logical program execution steps. Spatial localization is also hard, since any signal could be relevant at any point in time [4].

Waveform debugging in RTL design captures every signal's value at each clock cycle, but it is difficult to use for debugging hardware compiled from high-level languages. This is because [5]:

- Time is tracked in clock cycles rather than logical program steps, making **temporal bug localization** hard.
- All signals are always potentially relevant, complicating **spatial bug localization**.
- There is no clear mapping from waveform data to high-level source code, unlike software debuggers.

An Example for Waveform debugging is shown in Fig. 3.2. This waveform provides a clear example of debugging an APB (Advanced Peripheral Bus) interface through waveform analysis. It illustrates both write and read transactions and highlights the temporal relationships among key APB signals such as pclk, psel, pready, penable, pwrite, paddr, pwdata, and prdata. In the first visible transaction, the peripheral is selected (psel = 1), and with penable and pwrite asserted, a write operation is performed to address 0x10 with the data d99ce1a6. Later in the sequence, a read operation is shown where pwrite = 0, and once the slave asserts pready, valid data (599c9791) is read from address 0x2D. Notably, the waveform also captures instances where pready is deasserted, indicating that the slave is not ready to complete the transaction, which causes a delay. This is a crucial observation when debugging performance issues or protocol violations. By analyzing the sequence of these signal transitions, engineers can validate correct protocol behavior, ensure data integrity, and detect timing or handshaking issues. Conventional waveform debugging relies on manual signal tracing and human-in-the-loop efforts. However, LLM can be used for automated waveform analysis.

3.3.3 Code-Level Tracing/Debugging

Code-level tracing is widely applied in software and TLM development. It enables the tracing of function calls, variable changes, and program flow. While this technique is not typically applicable at the gate-level, it remains useful at the RTL and SoC levels where function-level insights into hardware design or high-level system interactions are needed. Code-level tracing helps developers pinpoint where things go wrong in the execution flow and can also be helpful in debugging more complex systems, where interaction between different components is key [6].

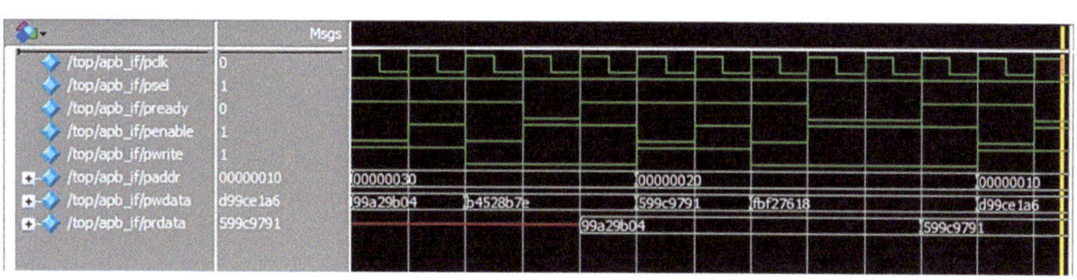

Fig. 3.2 An example for waveform debugging

Code-level tracing is a powerful debugging technique used extensively in software, TLM, and RTL verification environments. In hardware design, especially at the RTL and SoC levels, it enables engineers to trace function calls, observe variable changes, and monitor the flow of execution through structured logging. For example (Fig. 3.3), in a SystemVerilog UVM testbench, a typical driver class might include uvm_info statements to log the address, data, and operation type (read or write) of a transaction as it progresses through the run_phase. When the driver receives a transaction item, it logs the relevant details and then calls a function such as send_to_dut, which in turn logs further internal actions. These messages serve as checkpoints in the execution flow, allowing developers to understand the sequence of operations and detect where something might have gone wrong. This kind of tracing is especially valuable in complex SoC-level simulations where multiple components interact, and pinpointing issues in functional behavior requires visibility into the software-like control flow of the hardware verification environment. Unlike gate-level debugging, which relies on waveform-level signal tracing, code-level tracing provides high-level functional insight, enabling faster and more intuitive diagnosis of design or integration issues [7].

3.3.4 Printf-Style Debugging

To effectively troubleshoot and debug a program, using print statements is a valuable technique to gain a better understanding of its behavior. By strategically placing print statements, you can track the flow of execution and pinpoint where errors occur. For example, placing prints at the start and end of functions, as well as before and after loops, offers insight into the program's progress. Additionally, printing the values of key variables at different stages allows you to monitor data changes throughout the program. It's important to ensure that your messages are clear and informative to enhance the debugging process and make it more efficient.

```
// driver.sv
class ahb_driver extends uvm_driver #(ahb_seq_item);
  virtual task run_phase(uvm_phase phase);
    ahb_seq_item req;
    forever begin
      seq_item_port.get_next_item(req);
      `uvm_info("DRIVER", $sformatf("Got item: addr=%h, data=%h, write=%0b",
                    req.address, req.write_data, req.read_write), UVM_MEDIUM)
      send_to_dut(req);  // Function call trace point
      `uvm_info("DRIVER", "send_to_dut completed", UVM_LOW)
      seq_item_port.item_done();
    end
  endtask

  function void send_to_dut(ahb_seq_item req);
    // Simulate drive logic and log key steps
    `uvm_info("SEND", $sformatf("Driving transaction: AHB HADDR=%h", req.address), UVM_DEBUG)
    // ... drive logic here ...
  endfunction
endclass
```

Fig. 3.3 Code-level tracing example

Printf-style debugging is a common technique at the software and TLM levels. It is useful for tracking variable values, program flow, and checking the state of various conditions. At the RTL and gate-levels, printf-style debugging becomes impractical due to the low-level nature of these environments, where debugging must be more focused on signal and data flow. However, at the SoC and physical-level, engineers can use similar techniques to observe the system's high-level behavior, such as toggling ports or using LEDs to indicate status.

In a UVM-based testbench simulating an AHB transaction, printf-style debugging is used to track the behavior of the master driver during a write operation. As an example (Fig. 3.4), at simulation time 10 ns, the log message indicates that the master driver component (uvm_test_top.env.ahb_agent.mst_driver) is starting a transaction. The subsequent message prints the transaction details, including the address (0x8C79D918), data (0x61D7EE7A), transfer size (word—32 bits), burst type (SINGLE), and transaction type (NONSEQ). These details help verify that the correct transaction was generated by the driver. At 12 ns, another message confirms that the write data has been accepted by the DUT, signaling successful communication. Finally, at 15 ns, a log is printed indicating the end of the transaction. These timestamped messages, enriched with hierarchical component paths and contextual tags like [AHBDRV], allow engineers to trace and validate the flow of transactions. They are especially helpful for pinpointing issues in protocol handling or timing mismatches.

Dumping memory content is a widely used RTL debugging technique to gain visibility into the internal state of a design, particularly when debugging functional issues, incorrect data handling, or protocol mismatches. In many digital systems, internal memories such as FIFOs, SRAMs, register files, or cache structures hold critical data that cannot always be observed directly from output ports. By dumping memory content; either continuously during simulation or at specific breakpoints; engineers can inspect whether data is being written and read correctly, if addresses are valid, and whether data corruption or unexpected overwrites have occurred.

This technique is especially helpful when debugging state machines that interact with memory, or when verifying the behavior of custom memory controllers. Memory content can be dumped using $display, **$writememh**, or $writememb system tasks in Verilog, typically at the end of a simulation or during runtime under certain conditions. For example, one might trigger a memory dump when an error signal is asserted or after a test finish to validate that the memory contents match the expected golden model. In emulation or

```
UVM_INFO @ 10ns: uvm_test_top.env.ahb_agent.mst_driver [AHBDRV] Starting transaction
UVM_INFO @ 10ns: uvm_test_top.env.ahb_agent.mst_driver [AHBDRV] Sending WRITE transaction:
    Address  = 0x8C79D918
    Data     = 0x61D7EE7A
    HSize    = 0x2 (Word - 32 bits)
    HBurst   = SINGLE
    HTrans   = NONSEQ
UVM_INFO @ 12ns: uvm_test_top.env.ahb_agent.mst_driver [AHBDRV] Write data accepted by DUT
UVM_INFO @ 15ns: uvm_test_top.env.ahb_agent.mst_driver [AHBDRV] Ending transaction
```

Fig. 3.4 Printf-style debugging example

post-silicon environments, memory-mapped access or scan-based dump techniques may be used to retrieve memory content indirectly.

3.3.5 Assertion-Based Formal Debugging

3.3.5.1 Assertions for Pre-Silicon Debug

Assertions are widely used across all levels of debugging, from software through to post-silicon validation. In software, assertions help ensure that the program behaves as expected at runtime. At the TLM (Transaction-Level Modeling) and RTL (Register Transfer Level) levels, assertions are used to validate that the design logic adheres to functional specifications. At higher levels, such as SoC-level and physical-level, assertions check the overall system integrity, ensuring that the SoC operates as expected. Assertions are a key tool for detecting errors early in the design phase and can help catch issues in real-time systems. During FPGA testing, unexpected behavior may surface that cannot be reproduced in simulation. This discrepancy could stem from not knowing the precise input sequence that triggers the issue, or from the known inputs being too extensive to simulate practically. One approach is to directly assert or cover the behavior in question, essentially reverting to a testbench-driven method. This can be effective if the bug results from a rare but shallow input sequence requiring precise timing. If the issue occurs deep in execution and is unreachable from reset in a few cycles, using a simulation trace as a starting state can help—particularly in systems with a structured bring-up sequence.

To perform property checking in hardware systems, engineers utilize property specification languages such as PSL (Property Specification Language) and SVA (SystemVerilog Assertions) to formally describe system requirements. These specifications are then compared against a mathematical model of the design using model checking tools [8].

Assertion-Based Verification (ABV) has emerged as a highly effective technique in the hardware industry. By embedding assertions into the design, engineers can significantly improve the observability and controllability of the system. This leads to quicker error localization and reduced debugging time. Assertions can be applied across multiple abstraction and verification levels, including both pre-silicon validation and post-silicon debug. At higher abstraction levels, such as the Transaction-Level Model (TLM), assertions enable faster simulation and validation of complex systems. TLM, being more abstract than RTL, is well-suited for validating large designs and hardware/software co-design environments.

At the Register-Transfer Level (RTL), SystemVerilog Assertions (SVA) are widely used for IP verification to ensure that the design logic adheres to functional requirements. Assertions are typically derived from these requirements during the RTL coding phase. There are two main types of assertions [9]:

A. Immediate Assertions: These evaluate a condition instantaneously, without waiting for time-based events. They are typically placed inside always blocks and respond to signals defined in the sensitivity list (which can be level- or edge-triggered). If the specified condition is met, a pass message is displayed; otherwise, a failure message is shown. Immediate assertions are useful for checking state conditions in real-time. (See Listing 3.1 for an example.)

B. Concurrent Assertions: These assertions evaluate time-dependent behaviors by monitoring signal changes on clock edges. They consist of an antecedent (before the |-> implication operator) and a consequent (after the operator). If the antecedent is true, the assertion checks whether the consequent holds. If so, the assertion passes. Messages can be conditionally displayed, and severity levels can be assigned. The disable iff construct allows assertions to be temporarily deactivated under specific conditions. Assertions can be called to facilitate monitoring, debugging, or deactivation during runtime. (Refer to Listing 3.2 for an example.)

Listing 3.1 An Example for Immediate Assertions

```
always @(sensitivity_list)
  begin
  assert (condition) $display ("Pass");
          else $display("Fail");
  end
```

Listing 3.2 An Example for Concurrent Assertions

```
property PROPERTY_NAME;
          @(posedge clk) antecedent |-> Consequent;
  Endproperty
  assert property(PROPERTY_NAME;) $display("Pass")
  else $display("Fail");
```

SystemVerilog assertions can be executed in both simulation and formal verification environments. During simulation, assertions are evaluated against testbench-generated stimuli. In formal verification, specialized tools exhaustively explore all possible input sequences using internal engines, independent of testbench vectors. To guide and constrain these tools, engineers can use the assume keyword alongside assert and cover to express assumptions about inputs and behavior. Formal tools are language-agnostic as long as the hardware description can be mathematically represented: such as through binary decision diagrams. For instance, VHDL-based designs can also incorporate assertions using the bind construct to associate SVA logic within the simulation environment.

3.3.5.2 Synthesizable Assertions for Post-Silicon Debug

Since SystemVerilog Assertions (SVA) are written in a declarative language separate from the procedural RTL design language, synthesis tools typically ignore them during implementation. However, there is a growing interest in synthesizing assertions to enhance in-system observability. Achieving full visibility of an actively running design is critical for effective and rapid bug localization, yet remains a significant challenge due to the limited access to internal signal states and nodes.

To address this, various EDA tools have been developed to improve runtime observability. These tools usually rely on event-based triggers: once a trigger condition is met, the tool captures the current state of signals, registers, and memory, and displays this data as a waveform. While this method may be practical for small designs with limited complexity, it becomes unwieldy for large-scale systems such as SoCs that contain numerous IP blocks and signal paths; making comprehensive manual analysis infeasible.

Although recent tools aim to support assertion synthesis, few fully accommodate the complete range of SVA operators. Meanwhile, research into hardware-based assertion checkers has yielded a variety of implementation strategies, each with its own trade-offs.

Synthesizing assertions into hardware allows designers to compare actual behavior against intended functionality in real-time, which significantly simplifies bug localization; especially for elusive corner-case issues that traditional simulation may miss due to testbench limitations. Furthermore, reconstructing the post-silicon bug environment within a testbench using SVA properties can offer improved observability compared to post-silicon debugging alone, often providing clearer insight into the source of functional failures.

Many works introduce a novel architecture for a **SystemVerilog Assertions (SVA) synthesis compiler** that transforms non-synthesizable SVAs into synthesizable Verilog modules. The goal is to enable conversion into hardware logic circuits, which can monitor the behavior of a running design directly on FPGA hardware. The proposed architecture follows two core principles: a **simple compiler structure** and a **generic implementation** approach.

The motivation behind synthesizing assertions lies in their potential to greatly **enhance debugging efficiency**. When synthesized into hardware, these assertions can help detect and localize design errors in real time, offering full visibility into the internal behavior of the design on an FPGA. This makes the **prototyping and verification of complex SoCs** significantly more accurate and efficient, allowing engineers to pinpoint the exact cause of system failures. Moreover, such **hardware-based assertion checkers** can continuously monitor the performance of critical FPGA-based applications. As a result, **resource utilization** becomes a crucial factor, ensuring that the implementation remains area-efficient.

3.3.6 Simulation, Emulation and FPGA Prototyping

When developing and verifying digital designs, especially at the RTL level, simulation, emulation, and FPGA prototyping each offer distinct advantages in terms of debugging

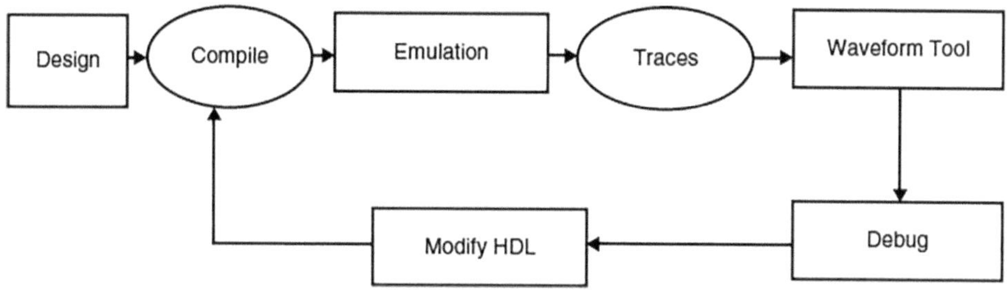

Fig. 3.5 Debugging flow at RTL level

capabilities, speed, and visibility (Fig. 3.5). A typical RTL-level debugging flow starts with simulation for early bug detection, progresses to emulation for full-system verification with manageable debug capabilities, and finally uses FPGA prototyping for performance validation and late-stage bug hunting. Designers typically cycle between these environments to isolate and fix complex issues efficiently.

By combining these methodologies, engineers can achieve comprehensive and scalable verification with appropriate debugging support at each stage of the design cycle.

By harnessing the unique strengths and capabilities of logic simulation, emulation, and FPGA prototyping, the chip design industry has established a comprehensive and robust verification framework that significantly enhances the overall chip development process. Logic simulation offers detailed and cycle-accurate analysis early in the design phase, enabling designers to detect and fix functional issues at a granular level. Emulation, on the other hand, provides accelerated hardware-level verification with real-time observability, making it ideal for validating complex system behaviors and running extensive test scenarios that are impractical in pure simulation. FPGA prototyping further complements these methods by allowing near-final hardware validation and software development in parallel with the chip design, thus shortening the development cycle and reducing time to market. Together, these methodologies form an integrated solution that not only accelerates design iterations but also guarantees high design accuracy and reliability.

As technology continues to evolve, heterogeneous computing architectures; comprising diverse computing units with distinct structural designs and data processing approaches; have become the dominant paradigm in modern chip design. This shift necessitates the adoption of tailored verification methodologies specific to the unique characteristics of each computing element. Recognizing the complexities and challenges posed by heterogeneous architectures, chip design companies have widely adopted a multi-faceted verification strategy, leveraging different verification tools and techniques at various stages of the design lifecycle. This strategic combination optimizes verification efficiency, ensuring thorough validation while simultaneously reducing development time. Such an approach has gained broad acceptance and is now a standard practice across numerous sectors within the semiconductor industry, reflecting a collective industry consensus to balance speed and accuracy in the pursuit of rapid innovation.

3.3.6.1 Simulation

Simulation is the most traditional and widely used method during early stages of design verification. It provides full visibility into internal signals, precise control over timing, and easy integration with assertions and checkers. Simulation tools support powerful debugging features such as waveform inspection, breakpoints, and code coverage. However, simulation is slow, especially for large or full-chip designs, which limits its use for long or complex test sequences [10]. Simulation can happen at different levels of abstraction. Register-Transfer Level (RTL) simulation plays a critical role in the functional verification and performance evaluation of digital circuits described in hardware description languages (HDLs) such as Verilog. It accurately models the behavior of signals and state elements such as registers across clock cycles. For example, in a two-stage pipeline that computes the dot product of two four-element vectors, an RTL simulator uses a testbench to provide inputs cycle by cycle, monitors outputs, and checks them against a reference model. Beyond functional verification, RTL simulation is also used to analyze performance metrics such as latency and estimate power consumption by tracking component activities. The focus is typically on synchronous digital designs, which dominate real-world applications, consisting of clocked registers and combinational logic. Although support for multiple clock domains is possible, many simulations assume a single clock for simplicity. Conceptually, RTL simulation can be abstracted as executing a dataflow graph where nodes represent logic operations and edges carry data values. Each simulation cycle evaluates these nodes based on data dependencies. This abstraction enables modern simulation tools such as Verilator, ESSENT, RepCut, and Cuttlesim to transform HDL code into intermediate representations (IRs) and further into efficient software models, often in C++. This translation takes advantage of the similarities between HDL and software semantics, simplifying operations like arithmetic. Two major optimization strategies for RTL simulation include parallelizing the execution of the dataflow graph and skipping evaluation of nodes with unchanged inputs. Despite these optimizations, current simulators face performance bottlenecks, revealing the need for further acceleration, possibly through hardware-assisted simulation techniques.

Designing modern chips with upwards of 10 billion transistors is becoming increasingly complex and costly, especially as Electronic Design Automation (EDA) tools struggle to keep up in terms of performance. This stagnation is particularly evident in Register-Transfer Level (RTL) simulation, a critical step for accurate functional modeling. For moderately sized blocks, ranging from 10 million to 100 million transistors, RTL simulation typically achieves only a few thousand cycles per second (CPS). To put this in perspective, simulating just 1 s of execution for a core operating at 1 GHz with one instruction per cycle could take more than 10 days. As chip designs scale, the inefficiency of RTL simulation becomes even more pronounced, making it impractical for early-stage development and limiting its use primarily to final verification stages [11].

3.3.6.2 Emulation

Emulation offers a middle ground between simulation and hardware prototyping. It runs the design on specialized emulation hardware and provides much faster execution than simulation, making it suitable for full-system verification. While visibility is more limited than in simulation, modern emulators include features like transaction-level tracing, waveform dumping, protocol checkers, and virtual I/Os. Emulation allows extensive debugging with reasonable visibility, although setting up and compiling designs for emulation can be time-consuming. Moreover, the emulation platform includes specialized verification intellectual properties (VIP) that offer essential verification interfaces. It can support a wide range of common interface protocols such as APB, AHB, AXI4, AXI4-Stream, AXI4-Lite, UART, SPI, I2C, DDR, Ethernet, USB, PCIe, SPI Flash, NAND Flash, and more. This comprehensive coverage addresses the majority of verification application needs. Additionally, S2C is capable of developing customized solutions tailored to specific customer requirements.

In-circuit emulation (ICE) is a technique used to test and debug microcontrollers and other embedded systems using an emulator that is physically connected to the target system. In this technique, the emulator is connected to the target system through a JTAG or other debug interface, and the code is executed on the target hardware. In-circuit emulation is particularly useful when the target hardware is available and when the developer needs to test the code in a real-world environment. This technique enables developers to observe the behavior of the target hardware and to identify and fix bugs that may only appear when the code is executed on the actual hardware (Fig. 3.6).

3.3.6.3 FPGA Prototyping

FPGA prototyping focuses on running the design on real hardware (FPGAs), delivering the fastest execution speed and near real-time performance, making it ideal for software development, system integration, and validation. However, its debugging capabilities are the most limited due to resource constraints and poor visibility into internal signals. Debugging often requires instrumenting the RTL with FPGA debugger logic (such as

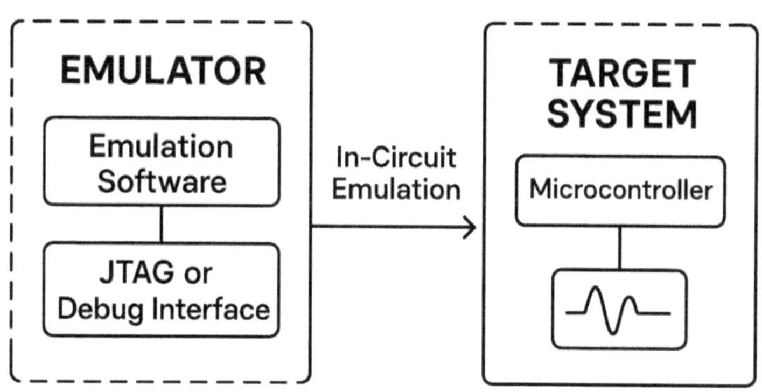

Fig. 3.6 In-circuit emulation (ICE) workflow for embedded system debugging

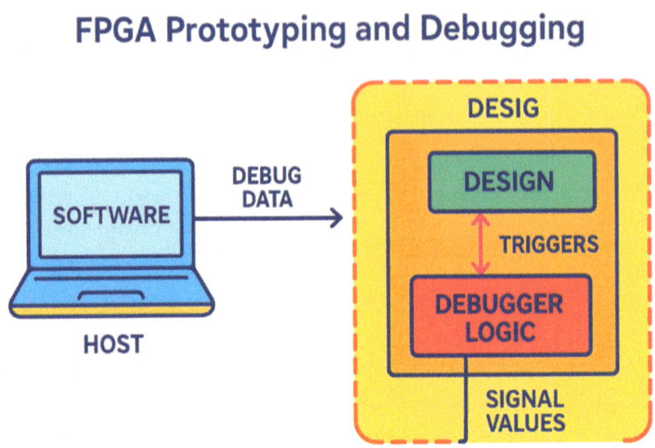

Fig. 3.7 FPGA prototyping workflow

Integrated Logic Analyzers like Xilinx's ILA or Intel's SignalTap) to probe internal signals. Because of limited on-chip memory and routing, designers must carefully choose which signals to observe.

To overcome the limited observability in FPGA-based prototyping, designers embed specialized **debugger logic blocks** in the FPGA fabric. These debugger blocks allow capturing signal values, triggering on specific conditions, and uploading captured data to the host.

These techniques help achieve a compromise between FPGA performance and visibility during debugging. However, they also require careful planning during the design compilation phase due to limitations in routing and area (Fig. 3.7).

3.3.6.4 A Comparative Comparison

In digital design verification, various techniques are used at different stages to ensure correctness and performance of the design. RTL Simulation is an early-stage, fast, and flexible approach that verifies the design's functionality without timing details. Gate-Level Simulation comes next, verifying the synthesized netlist with detailed timing information to ensure timing correctness, though it is slower and more resource-intensive. Emulation uses specialized hardware platforms to run the design with very high timing accuracy and extensive debug capabilities, making it ideal for late-stage, full system validation of complex designs. FPGA Prototyping maps the design onto reconfigurable FPGA hardware for real-time execution and early hardware/software integration testing. It offers faster iteration than emulation but with less debug visibility and approximate timing accuracy (Table 3.2) [12].

Table 3.2 A comparative comparison between simulation, emulation and FPGA prototyping

Feature	RTL simulation	Gate-level simulation	Emulation	FPGA prototyping
Verification stage	Early functional verification	Post-synthesis timing and logic check	Late-stage full system validation	Early hardware/ software integration
Speed	Fast	Slow	Slower than prototyping, but faster than simulation for large designs	Fast, real-time execution
Timing accuracy	No timing details	Detailed timing with delays	Very accurate timing and behavior	Approximate timing (FPGA hardware limited)
Debug capability	High visibility, easy waveform debug	High visibility, waveform debug	Extensive debug and visibility tools	Limited debug visibility
Flexibility	High, easy to modify and rerun	Medium, changes require resynthesis	Low, specialized hardware setup needed	Moderate, requires FPGA place and route
Cost	Low (software only)	Medium (compute and licenses)	High (specialized emulator hardware and software)	Medium (FPGA boards and tools)
Use cases	Functional correctness and design validation	Timing closure and gate-level checks	Complex SoC validation, full system debug, software bring-up	Hardware/software co-verification, early system validation

3.3.7 Linting Tools: Static Debugging

Linting tools are an essential part of early-stage debugging, especially for detecting common coding errors, potential bugs, and violations of coding standards before synthesis or simulation. These tools analyze source code statically—without running it—to identify issues such as undeclared variables, unreachable code, race conditions, or inconsistent reset behaviors. Linting helps enforce consistency, catch subtle mistakes early, and improve code readability and maintainability. While it cannot replace dynamic simulation or emulation, linting acts as a low-effort, high-reward step to eliminate trivial issues that could otherwise propagate into more complex and time-consuming debug sessions.

With the growing complexity of modem circuit designs and their corresponding RTL code, linting has emerged as a crucial first step within the broader scope of formal verification methodologies. Formal verification generally comprises two widely used techniques: Formal Equivalence Checking (EC) and Formal Property or Model Checking (MC). Linting is typically positioned within the model checking branch.

Formal Equivalence Checking (EC) is used to validate the functional correctness of a design by comparing it to a golden reference model using testbenches and stimulus vectors. On the other hand, Formal Model Checking (MC) is employed to examine design specifications and identify potential root causes of errors that may occur during the ASIC development process.

Model checking can be viewed as a hierarchical process composed of four distinct layers based on design complexity, application scope, and the potential benefits they provide (as illustrated in Fig. 3.5). The foundational layer, referred to as Automatic Formal, includes a variety of linting tools—this paper focuses on this layer. These tools are primarily used to detect design issues such as pragma violations and unreachable (dead) code.

The second layer, Formal Apps, targets specific verification challenges such as power optimization, connectivity checks, and clock domain crossing issues. The third layer, Assertion-Based Verification (ABV), enables engineers to create and validate assertions that capture the intended behavior of RTL code. Finally, the fourth and most comprehensive layer, End-to-End Formal, is designed to verify the full functionality of design blocks or systems under test.

Design verification is critical in the digital design lifecycle, typically consuming around 80% of the overall project timeline. This underscores the importance of reducing verification effort wherever possible. The aim of this work is to introduce a tool that automatically detects and addresses the most critical RTL design issues, thereby shortening the verification phase in industrial digital design workflows.

Conventional linting tools often generate lengthy reports filled with errors and warnings, many of which must be resolved manually: placing a significant burden on verification engineers and leading to time inefficiencies. However, several of these issues are predictable and can be automatically diagnosed and corrected. The proposed tool addresses common timing violations and mismatches between simulation and synthesis, offering automated fixes. It has been tested on Verilog and SystemVerilog source code, which are prevalent hardware description languages in the semiconductor industry. Table 3.3 summarizes RTL Issues that can be discovered by linting tools.

3.3.8 Logs-Based/Transcript-Based Debugging

Logs play a critical role in both software and hardware debugging by providing a chronological record of events, values, and execution paths. In software, logs may include messages from printf statements, debug macros, or logging frameworks that print variable states and execution checkpoints. In hardware emulation or simulation, logs can be generated through monitors, assertions, or custom print tasks. Well-structured logs help engineers trace system behavior, correlate it with stimulus, and identify unexpected transitions or values. Logging is particularly useful when dealing with long-running simulations or complex SoC-level interactions where replaying the entire scenario for every inspection is impractical. Figure 3.8 shows an example of log-based debugging.

Table 3.3 RTL issues that can be discovered by linting tools

Issue	Description	Linting tool action
Incomplete sensitivity list	Missing signals in the always block sensitivity list.	Detects missing signals in combinational blocks and replaces list with (*) for synthesis compatibility.
Unused signals/ ports	Signals or ports declared but not assigned to any logic.	Statically removes unassigned signals or ports before simulation.
Glitch from gated clock	Gated clocks implemented via poor logic (e.g., assign clk = enable && clk) can cause glitches.	Adds intermediate latched signal to synchronize enable with clock.
Missed else statement	An if condition without a matching else leads to latch inference.	Detects and pinpoints exact missing else location for correction.
Non-full case statement	Case block missing a default, allowing latch inference for unhandled cases.	Detects incomplete case and identifies position for adding default.
Partially assigned outputs	Only some conditions assign a value to an output signal.	Flags signals not assigned in all branches and prompts full assignment coverage.
Conditional assignment outside reset	Sequential logic missing full reset coverage.	Warns if not all paths assign a value during reset phase in always_ff or always @(posedge clk) blocks.
Improper use of tri-state buffers	Misuse or partial control of tri-state enables can infer latches or undefined values.	Detects missing control paths and prompts review of tri-state logic.
Mixed clock domains	Assignment of signal across asynchronous clock domains without synchronization.	Flags cross-domain signals without synchronizers; prevents metastability that can mimic latch behavior.
Latch in FSM encoding	FSM encoding style (e.g., no default state or missing state transitions) can create latches.	Ensure FSM always assigns the next state in all possible conditions.

```
# UVM_INFO AHB_mdriver.sv(104) @ 130: uvm_test_top.env_h.master_agent_h.mdriver_h [ahb_mdriver] Transaction From AHB Master Driver..
# -------------------------------------------------------------------------------------------
# Name                        Type                Size  Value
# -------------------------------------------------------------------------------------------
# req                         ahb_mxtn            -     @1471
#   reset                     integral            1     'h1
#   trans_type                array(transfer_t)   2     -
#     [0]                     transfer_t          2     NONSEQ
#     [1]                     transfer_t          2     BUSY
#   address                   da(integral)        1     -
#     [0]                     integral            32    'h8c79d918
#   trans_size                size_t              3     HALFWORD
#   burst_mode                burst_t             3     INCR
#   read_write                rw_t                1     WRITE
#   write_data                da(integral)        1     -
#     [0]                     integral            32    'h617dee7a
```

Fig. 3.8 An example of log-based debugging

The image provided is a snapshot of a UVM log message generated during simulation, specifically from the AHB master driver component in a SystemVerilog-based verification environment. This log represents a typical example of a **log-based debugging technique**, which is a fundamental approach in hardware verification to trace, analyze, and debug stimulus or behavior of a testbench or DUT (Design Under Test).

In this instance, the message originates from line 104 of the AHB_mdriver.sv file and is printed using the UVM_INFO macro. It captures the details of an AHB transaction initiated by the master driver. The transaction object req., of type ahb_mtxn, encapsulates all the relevant fields of the AHB transfer. The log reveals several important attributes of the transaction. For instance, the reset signal is asserted (1'h1), which is crucial because initiating a transaction while reset is active might indicate unintended behavior. The trans_type array shows two transfer types: NONSEQ and BUSY, helping verify that the protocol's sequence rules are correctly followed. The address is logged as 32'h8c79d918, ensuring the master targets the correct memory region. Other fields include trans_size (HALFWORD), which specifies the data size being transferred, burst_mode (INCR) indicating an incrementing burst operation, read_write (WRITE) showing the direction of data transfer, and write_data (32'h61d7ee7a), which is the actual data being written.

By logging this level of detail, verification engineers can trace issues back to specific transactions, validate stimulus generation against expected behavior, and quickly identify incorrect or unintended sequences. This kind of log output is instrumental in root cause analysis during regression failures and in debugging protocol violations, making log-based debugging an indispensable technique in the UVM methodology [13, 14].

3.3.9 Triggers-Based Debugging

Triggers are specialized mechanisms used to halt or capture information when a particular condition is met. In simulation, triggers can stop the simulation when assertions fail, illegal states are entered, or signal values reach a certain threshold. In emulation or FPGA prototyping, triggers can be configured to capture a snapshot of the system—waveforms, registers, or memories—when the condition is met, such as a protocol error or unexpected signal toggle. They help isolate the root cause of bugs by narrowing down the time window and operational context in which the issue first manifests. Efficient use of triggers reduces the time spent shifting through irrelevant data and improves the debugging turnaround.

Consider a scenario where you're debugging a UART module that occasionally experiences data corruption on the receive path. In RTL simulation, you can use trigger-based debugging by setting up an assertion that fires when the receive FIFO (rx_fifo) is full and new data is still being written: an indication of an overflow condition (Fig. 3.9). When this

```
always @(posedge clk) begin
  if (rx_fifo_full && rx_data_valid) begin
    $display("Trigger: RX FIFO overflow detected at time %0t", $time);
    $stop; // Stop simulation to inspect waveforms
  end
end
```

Fig. 3.9 An example of trigger

trigger condition is met, the simulation halts, and a snapshot of waveform data around the event is captured, helping you isolate the issue to a flaw in the write-enable logic. In contrast, during emulation or FPGA prototyping, you cannot easily stop execution, but you can configure the system to automatically capture a trace window when the same trigger condition occurs. For instance, the emulator can log signal transitions, CPU register values, and memory states when the FIFO overflow happens, allowing you to analyze the software and hardware behavior leading to the issue. This targeted capture avoids the need to sift through large, irrelevant data logs and significantly improves debug efficiency, especially for intermittent or complex bugs.

3.3.10 Monitors

Monitors are passive verification components used to observe and validate the behavior of a design without interfering with it. They are commonly used in testbenches to check protocol adherence, signal transactions, or timing constraints. Monitors can be simple signal checkers or complex modules tracking interactions across interfaces such as AXI, PCIe, or AMBA protocols. They provide valuable coverage information and can be configured to flag anomalies or log critical events. When integrated with **scoreboards** and reference models, monitors also contribute to functional verification by ensuring that the design behaves as expected in various scenarios. Their non-intrusive nature makes them indispensable in both simulation and emulation-based debug workflows.

For example, consider a design incorporating an AXI interface. A monitor can be implemented in the testbench to passively observe AXI transactions such as read and write bursts, address alignment, and handshake signals like AWVALID, WVALID, and BREADY. This monitor does not drive or alter any signals but instead logs the sequence and timing of transactions to check protocol compliance. If it detects out-of-order responses or unexpected burst lengths, it can raise flags or log the events for later analysis. In more advanced setups, the monitor may compare observed transactions against a reference model to validate the correctness of data transfers. This makes monitors especially valuable during functional simulation and emulation, where visibility into internal signals is limited. Their ability to provide continuous, passive observation without affecting the system's behavior makes them essential for verifying complex protocols and ensuring robust system integration (Fig. 3.10).

Scoreboarding is a powerful technique in functional verification used to validate the behavior of an RTL design against a **golden reference model**. The scoreboard acts as a dynamic checker: it records expected outputs based on the inputs sent to the design (often using a high-level algorithm or functional model) and then compares those expectations with the actual outputs received from the DUT (Design Under Test). This decouples input generation from output checking, allowing tests to focus on stimulus without embedding correctness checks in the same block. For example, in an AXI protocol-based design, the scoreboard would track outstanding transactions—recording each write or read command;

```
class axi_write_monitor;
  // Clock and reset
  logic clk;
  logic rst_n;

  // Observed AXI signals
  logic        AWVALID;
  logic [31:0] AWADDR;
  logic [7:0]  AWLEN;
  logic        AWREADY;

  logic        WVALID;
  logic [31:0] WDATA;
  logic        WREADY;

  logic        BVALID;
  logic [1:0]  BRESP;
  logic        BREADY;

  // Constructor
  function new(input logic clk_i, input logic rst_ni);
    clk   = clk_i;
    rst_n = rst_ni;
  endfunction

  // Monitor main process
  task run();
    forever @(posedge clk) begin
      if (!rst_n) begin
        // Wait for reset deassertion
        continue;
      end

      // Detect write address channel activity
      if (AWVALID && AWREADY) begin
        $display("[AXI Monitor] Write Address: 0x%08x, Burst Length: %0d @ %0t", AWADDR, AWLEN, $time);

        // Check address alignment (assuming 4-byte word alignment)
        if (AWADDR[1:0] != 2'b00) begin
          $error("[AXI Monitor] Misaligned write address detected: 0x%08x @ %0t", AWADDR, $time);
        end

        // Check for illegal burst lengths
        if (AWLEN > 8'd15) begin
          $warning("[AXI Monitor] Unusually long burst length: %0d @ %0t", AWLEN, $time);
        end
      end

      // Detect write data channel activity
      if (WVALID && WREADY) begin
        $display("[AXI Monitor] Write Data observed: 0x%08x @ %0t", WDATA, $time);
      end

      // Detect write response
      if (BVALID && BREADY) begin
        if (BRESP != 2'b00) begin
          $error("[AXI Monitor] Write response error: BRESP = %b @ %0t", BRESP, $time);
        end else begin
          $display("[AXI Monitor] Write response OK @ %0t", $time);
        end
      end
    end
  endtask
endclass
```

Fig. 3.10 An example of monitor

and verify whether the DUT returns the correct data, in order and with expected latency or behavior. If discrepancies are detected, the scoreboard logs detailed mismatches, making it easier to trace the root cause. This method ensures systematic, repeatable validation, especially effective for complex protocols, pipelines, or memory coherency checks, and is often paired with UVM-based environments.

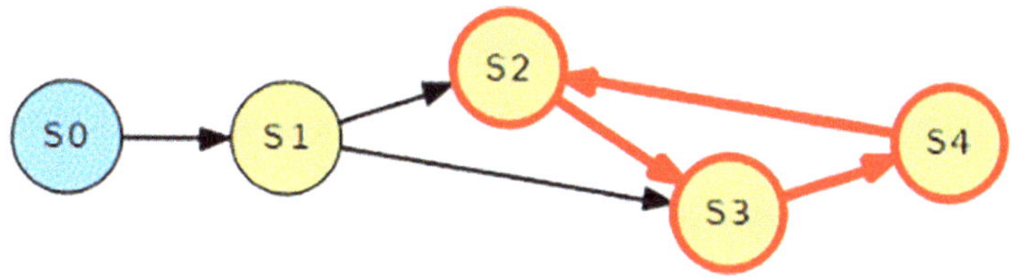

Fig. 3.11 An example for FSM

3.3.11 FSM Debugging

Finite State Machine (FSM) debugging is particularly relevant for the RTL, gate-level, and SoC levels (Fig. 3.11). At these levels, the design's state machine transitions are critical, and FSM debugging techniques allow engineers to observe and validate state changes, identify deadlocks, or find incorrect state transitions. FSM debugging is less useful at the higher levels (software and TLM), as the focus is on design verification and functional behavior rather than the state transitions themselves.

3.3.12 Clock/Reset Issues

Issues related to clocks and resets are critical in hardware design, especially in the **RTL and gate-level design phases**. Clock skew, improper reset behavior, and clock domain crossing are common problems encountered at these stages. At the SoC level, engineers also need to check clock interactions between different subsystems, ensuring synchronization across the chip. At the physical-level, clock and reset issues may become even more complex and can cause timing violations if not properly managed. These issues are often diagnosed using tools such as simulators or waveform analysis. clocks debug happens at run-time. For example, clock is always 0 or not toggling.

3.3.13 Backdoor Access

Backdoor access refers to techniques that allow verification of environments or testbenches to directly read from or write to internal design registers, memories, or signals without going through the standard input/output interfaces of the DUT. This is especially useful for initializing memory contents, injecting values into internal states, or observing data that might not be accessible through regular ports. Backdoor access can be implemented using mechanisms like **PLI/VPI (Programming Language Interface/Verilog Procedural Interface)** or **DPI (Direct Programming Interface)**, which allow C/C++ code to interact with SystemVerilog simulations. For instance, during bring-up or post-reset verification, a testbench might use DPI to preload firmware into a memory array or to compare register values directly after a configuration sequence. Similarly, in low-level debugging, engineers can force signal values to specific states or extract internal waveforms without altering the DUT's functional paths. Backdoor techniques accelerate debug and enhance observability, especially when combined with UVM register models or emulation platforms where traditional read/write cycles would be too slow or complex.

3.3.14 Loopback Mode Technique

Loopback mode is a practical and powerful RTL debug technique used to isolate and validate internal data paths, interfaces, or protocol logic by routing outputs back into inputs, effectively creating a closed-loop system. This approach helps verify functionality without needing a full external environment and is especially useful in early design stages or when debugging communication interfaces such as UART, SPI, Ethernet, or even custom interconnects.

In loopback mode, the output of a module (e.g., transmit path) is connected internally to its corresponding input (e.g., receive path). This allows engineers to test end-to-end data flow, timing alignment, and protocol correctness without relying on external stimuli. For example, in a UART module, enabling loopback means transmitted data immediately appears at the receiver, allowing verification of serialization, framing, and deserialization logic.

Loopback can be implemented either in the RTL (via conditional logic or a configuration register) or in the testbench by forcing signals to connect internally. It's often controlled by a dedicated loopback_en signal. This mode is also useful for debugging FIFOs, DMA engines, or interconnect fabric, where validating write-read cycles within a single module can eliminate external sources of error.

By reducing dependencies on external test drivers or stimulus generators, loopback mode simplifies test scenarios and helps narrow down bugs to specific functional blocks. It's a reliable first step in interface validation and a valuable tool for identifying mismatches, delays, or data corruption within a self-contained setup.

3.4 Gate-Level Debugging Techniques: Low-Level

Gate-level debugging is a critical phase in the hardware design lifecycle, focusing on the detailed behavior of synthesized logic and physical interconnects. Key techniques such as netlist probing enable engineers to inspect connectivity and detect structural errors at the gate or physical level. Timing violation debugging targets setup and hold violations, leveraging both simulation and post-silicon validation to ensure timing correctness under real operating conditions. Gating and glitch analysis helps identify signal integrity issues and unintended switching activity, which may affect power or logic reliability. Protocol-aware debugging ensures that inter-module communication adheres to expected specifications, especially at RTL and SoC levels, while protocol analyzer devices play a vital role in post-silicon validation by capturing and analyzing real-time hardware transactions. Together, these techniques form a comprehensive toolkit for low-level debugging, bridging the gap between logical correctness and physical implementation to ensure robust, functional silicon [15, 16].

3.4.1 Netlist Probing

Netlist probing is a technique used primarily at the **gate-level** and physical-level of design. It involves inspecting the connections between logic gates or physical components. At these levels, netlist probing can help detect issues like improper connections, logic errors, or signal integrity problems that could lead to system failures. This technique is not applicable at higher software or RTL levels, where the focus is more on logical and functional design rather than physical implementation.

3.4.2 Timing Violation Debugging

Timing violations, such as setup and hold time violations, are a major concern at the **gate-level** and RTL stages. These violations are diagnosed by checking the timing constraints during simulation or post-silicon validation. At higher levels, such as SoC and physical-level, timing analysis tools can help detect violations across different system components. Post-silicon-level debugging also involves checking the real-world timing of signals to verify that the design meets its performance targets under actual operating conditions.

3.4.3 Gating and Glitch Analysis

Gating and glitch analysis are used primarily at the **gate-level** and physical-level to detect issues related to signal integrity and the improper switching of gates. These problems can lead to unwanted behaviors such as glitches or power consumption spikes. Engineers at the RTL and SoC levels may also use this technique to ensure that the clock gating or other control mechanisms function properly before implementation. At the physical-level, tools like oscilloscopes are often used to observe and analyze these glitches in real-time.

3.4.4 Protocol-Aware Debugging

Protocol-aware debugging is crucial in system designs where multiple subsystems communicate through predefined protocols. It is useful at the TLM, RTL, and SoC levels, where engineers must ensure that data exchanges adhere to the expected protocol specifications. In high-level designs, such as TLM, this can help validate communication between abstract models. However, it is not typically applied at the physical-level, where hardware signals are already abstracted, and physical protocol analyzers are used instead.

3.4.5 Protocol Analyzer Device

A protocol analyzer/tracer device is a tool used primarily at the SoC and post-silicon levels. It helps engineers monitor and analyze communication protocols between different subsystems or between the chip and external devices. At lower levels, like the TLM and RTL, protocol analyzers are less useful, as the focus is more on logical validation and simulation. However, at the **physical-level**, these analyzers become essential for debugging real-world communication between hardware and external components, often used during post-silicon validation to check signal integrity and protocol adherence.

3.5 Debugging Challenges

Debugging in hardware and embedded system design is inherently complex due to multiple abstraction levels, tight hardware-software coupling, and limited visibility into internal system states. This section explores the main challenges engineers face, categorized into logical subsections for clarity [17, 18].

3.5.1 Hardware vs Software: What's Broken?

One of the first questions a designer or verification engineer must ask when encountering a malfunctioning system is: Is the issue rooted in hardware or software? This distinction is rarely obvious. For instance, a peripheral not responding correctly might stem from incorrect register logic in the RTL, or it could be caused by a misconfigured driver or faulty firmware sequence.

Such ambiguity is exacerbated during early SoC bring-up phases when both the hardware and software components are immature. Limited observability into internal signals: especially in post-silicon environments further complicates root cause analysis. As a result, coordinated instrumentation on both hardware (e.g., signal probes, logic analyzers) and software (e.g., logging, toggling GPIOs) is essential for effective debug.

3.5.2 Debugging at Different Abstraction Levels

Debugging behavior differs significantly depending on the abstraction level:

- **TLM-Level Debugging:** At the transaction-level modeling (TLM) abstraction, debugging is relatively straightforward. The high level of abstract abstracts away timing details, resulting in faster simulations and simpler analysis. Errors at this stage often pertain to functional logic, such as missing transactions or protocol mismatches, which are easier to trace using software-style debugging techniques like print statements and log analysis.
- **RTL-Level Debugging:** In contrast, debugging at the RTL level introduces many complexities. Engineers must account for clocking issues, reset sequencing, finite-state machine (FSM) behavior, and signal races. Signal visibility becomes crucial, often requiring the use of waveform viewers, assertions, and trace buffers. Timing-related bugs, glitches, and hardware protocol violations typically appear at this stage and are much harder to localize and reproduce.

3.5.3 Debugging UVM Macros and Infrastructure

The Universal Verification Methodology (UVM) is a widely adopted framework in RTL verification, offering structured, reusable testbenches. However, its complexity introduces new challenges:

- UVM macros (e.g., uvm_info, uvm_error, uvm_component_utils) encapsulate functionality that can obscure the flow of execution.
- Component factories and configuration databases can dynamically alter testbench behavior in ways that are hard to trace.
- Debugging phase sequencing, callbacks, and register models can be unintuitive without careful logging and naming conventions.

This abstraction, while powerful, often leads to situations where the root cause of a bug is hidden behind several layers of object-oriented infrastructure, making straightforward debugging difficult.

3.5.4 Limited Observability and Instrumentation

Effective debugging depends on what you can see and control. At the RTL and post-silicon levels, designers often face severe observability constraints:

- Internal nodes and signals are not always accessible without waveform dumps or embedded logic analyzers.
- Post-silicon debugging is particularly restricted, relying on pre-synthesized debug ports, status LEDs, or JTAG-based tools.

- Without assertion-based verification or runtime trace capture, identifying complex corner-case bugs becomes nearly impossible.

Moreover, emulation and prototyping platforms further limit what internal data can be monitored without affecting performance or timing.

3.5.5 Diagnosing Test Failures: Hardware, Software, or Environment?

When a test consistently fails on one hardware setup but passes on another, it typically points to a hardware issue. Conversely, if the same test fails consistently across multiple different hardware platforms, the problem is more likely rooted in the software. However, if the test results vary inconsistently or appear random on the same setup, the underlying cause is often environmental such as overloaded servers, unstable network connections, or other transient system conditions.

References

1. Dehbashi, M., & Görschwin Fey. (2014). *Debug automation from pre-silicon to post-silicon.* Springer.
2. Ziad, M. T., Ibn, Y. A., Watheq El-Kharashi, M., Salah, K., & AbdelSalam, M. (2015). Accelerating electromagnetic simulations: A hardware emulation approach. In *2015 IEEE international conference on electronics, circuits, and systems (ICECS)* (pp. 592–595). IEEE.
3. Mohamed, K. S. (2024). HW/SW co-verification and co-debugging. In *Heterogeneous SoC design and verification: HW/SW co-exploration, co-design, co-verification and co-debugging* (pp. 133–151). Springer Nature Switzerland.
4. Pit-Claudel, C., Bourgeat, T., Lau, S., & Chlipala, A. (2021). Effective simulation and debugging for a high-level hardware language using software compilers. In *ACM International Conference on Architectural Support for Programming Languages and Operating Systems (ASPLOS).* ACM.
5. Berlstein, G., Nigam, R., Gyurgyik, C., & Sampson, A. (2023). Stepwise debugging for hardware accelerators. In *Proceedings of the 28th ACM International Conference on Architectural Support for Programming Languages and Operating Systems, Volume 2 (ASPLOS '23), March 25–29, 2023, Vancouver, BC, Canada.* ACM.
6. Zhang, K., Asgar, Z., & Horowitz, M. (2022). Bringing source-level debugging frameworks to hardware generators. In *Proceedings of the 59th ACM/IEEE Design Automation Conference, DAC '22.* IEEE.
7. Salah, K. (2024). Revisiting UVM. In *2024 IEEE East-West Design & Test Symposium (EWDTS).* IEEE.
8. Mohamed, K. S. (2024). Pre-silicon verification and post-silicon validation methodologies. In *Heterogeneous SoC design and verification: HW/SW co-exploration, co-design, co-verification and co-debugging* (pp. 85–131). Springer Nature Switzerland.
9. Foster, H. D., Krolnik, A. C., & Lacey, D. J. (2004). *Assertion-based design.* Springer Science & Business Media.
10. Elsabbagh, F., Sheikhha, S., Ying, V., Nguyen, Q., Emer, J., & Sanchez, D. (2023). Accelerating RTL simulation with hardware-software co-design. In *MICRO '23: Proceedings of the 56th*

Annual IEEE/ACM International Symposium on Microarchitecture (pp. 153–166). ACM. https://doi.org/10.1145/3613424.3614257

11. López-Paradís, G., Li, B., Armejach, A., Wallentowitz, S., Moretó, M., & Balkind, J. (2023). Fast behavioural RTL simulation of 10B transistor SoC designs with Metro-Mpi. In *2023 Design, Automation & Test in Europe Conference & Exhibition (DATE), Antwerp, Belgium* (pp. 1–6). IEEE. https://doi.org/10.23919/DATE56975.2023.10137080
12. Retrieved from https://www.electronicsforu.com/electronics-projects/electronics-design-guides/fpga-prototyping-techniques
13. Retrieved from https://developer.arm.com/documentation/ihi0033/latest/
14. Salah, K. (2014). A UVM-based smart functional verification platform: Concepts, pros, cons, and opportunities. In *2014 9th International Design and Test symposium (IDT)*. IEEE.
15. Piper, L., & Vimjam, V. (2011). X-propagation woes: Masking bugs at RTL and unnecessary debug at the netlist. In *DVCon Proceedings*. Real Intent Inc.
16. Lee, S.-Y., Riener, H., & De Micheli, G. (2022). An automated testing and debugging toolkit for gate-level logic synthesis applications. *arXiv:2207.13487.*
17. Lettnin, D., & Winterholer, M. (Eds.). (2017). *Embedded software verification and debugging*. Springer.
18. Roychoudhury, A., et al. (2009). *Design for debugging and validation of embedded systems.* Elsevier.

AI-Driven Methods for RTL Debugging: Localization and Auto-Corrections of RTL Bugs "Towards Smart Debugging"

4

4.1 Introduction

In the rapidly evolving field of digital design, especially for complex systems-on-chip (SoCs), debugging RTL (Register Transfer Level) code has become one of the most critical and time-consuming tasks for engineers. As designs grow in size and complexity, traditional debugging techniques such as waveform analysis and manual inspection often fall short in terms of speed and accuracy. This is where AI-driven RTL debugging comes into play, offering solutions for faster bug localization and automatic correction, ultimately improving productivity and design quality.

AI-driven tools, leveraging machine learning (ML) algorithms and deep learning (DL) models, have made their mark in optimizing the RTL debugging process. These AI systems can analyze massive volumes of simulation and synthesis data far more efficiently than humans. By learning from previous designs and patterns, AI tools can identify anomalies, predict possible errors, and even suggest potential fixes based on historical data and design characteristics. It can collect the most used signals and add it to **indicator logic** for debug.

Localization is the process of identifying the exact location of a bug in an RTL design. AI tools can assist in this step by scanning through the design code, simulations, and waveforms to find patterns associated with common errors. AI localizes RTL bugs through a combination of pattern recognition, trace analysis, and control flow analysis. By training on large datasets of known RTL bugs and their corresponding fixes, AI models learn to recognize recurring patterns and apply this knowledge to new designs, enabling them to pinpoint likely bug locations. Trace analysis allows AI to compare actual signal transitions against expected behavior, identifying mismatches and flagging code regions where timing errors or incorrect signal driving may have occurred. Additionally, control flow analysis enables AI tools to scrutinize the behavior of finite state machines (FSMs) and control

K. Mohammed, *The Art of RTL Debugging*, Synthesis Lectures on Digital Circuits & Systems, https://doi.org/10.1007/978-3-032-15880-2_4

logic, detecting unreachable states, incorrect transitions, or dead code. For example, in a scenario where an FSM fails to transition correctly due to a misinterpreted input condition, the AI can detect that the transition rule is not activating and highlight the faulty input check, directing the designer to the exact line of code responsible for the issue.

Once the bug is localized, the next logical step is **auto-correction**. AI systems can go beyond just identifying problems; they can suggest or even implement fixes based on past solutions and predefined guidelines. AI auto-corrects RTL bugs by leveraging automated bug fix suggestions, code refactoring, and optimization algorithms. When a known issue like a timing violation is detected, AI-driven debugging tools can recommend corrective actions based on prior similar fixes: such as adjusting clock constraints or modifying logic to improve timing closure. In cases of structural issues like deeply nested conditions or poorly modularized code, the AI can suggest refactoring the RTL to enhance readability, synthesis efficiency, and maintainability. Furthermore, AI employs optimization algorithms to propose improvements, such as rebalancing pipeline stages, adjusting memory accesses to reduce contention, or restructuring clock domain crossings to mitigate metastability risks. For example, if a setup violation arises due to inadequate hold time, the AI can automatically insert a timing-friendly register or refine clock constraints, resolving the issue without requiring manual debugging or RTL changes from the designer.

AI-driven RTL debugging offers significant benefits, including faster debugging, higher accuracy, improved design quality, and reduced human error. By quickly analyzing complex signal behaviors and design traces, AI tools drastically cut down the time needed to identify and fix RTL issues, accelerating the overall design cycle. Leveraging large datasets of known issues and fixes, AI can detect subtle or non-obvious bugs that might be missed by human engineers, thereby increasing accuracy and reducing the risk of undetected flaws. Additionally, AI can automatically suggest or apply best practices, correcting inefficient coding styles and enhancing the reliability and maintainability of the RTL. By automating repetitive and error-prone debugging tasks, AI also reduces the potential for human mistakes and allows engineers to focus on higher-level design challenges [1].

Traditional RTL emulation offers high-speed simulation but is limited by poor observability and requires time-consuming recompilation for debugging internal signal issues. The proposed methodology in [2] allows designers to change internal signal values at runtime without re-synthesizing the design, using auto-generated **glue logic and multiplexers inserted into the RTL**. This empowers a real-time debug environment similar to AI-guided frameworks that rely on dynamic data collection and trace analysis. When integrated with AI tools, such scan-chain approaches could significantly boost the feedback loop for auto-correction, enabling pattern-based identification of misbehaving signals and faster validation of fixes. Although this method adds slight overhead area and increases compilation time, it drastically enhances runtime controllability and observability, which are critical for enabling intelligent, data-driven RTL debugging solutions.

The proposed work in [3] addresses the limitations of traditional simulation-based methods like BMC and ATPG, which detect but do not localize or correct bugs. By applying a **mutation-based** technique, it systematically generates and evaluates candidate RTL modifications to identify likely bug locations and propose corrections with minimal human intervention. This process mimics AI reasoning by learning from the testbench behavior and selecting mutations that restore correct functionality, making it a promising direction for integrating intelligent automation into RTL debugging workflows.

The proposed RTL bug localization methodology in [4] employs a **coverage-driven**, test-based statistical approach to automatically pinpoint faulty statements in HDL designs by analyzing the execution profiles of passing and failing testcases. After running a complete regression suite to achieve full coverage, the framework extracts which statements are exercised by each test and calculates a *suspicious score* for every RTL statement; defined as the ratio of failing tests executing it to passing tests. Statements heavily associated with failures and rarely seen in passing tests are ranked higher as probable bug sources. This ranking enables a focused and automated localization of errors without manual code inspection. Experimental results demonstrate the technique's effectiveness, accurately localizing bugs in complex RTL designs (over 5000 lines) within minutes; dramatically reducing debug time compared to traditional manual efforts that can take hours. The methodology assumes a single bug per run and depends on high-quality testcases for reliable bug detection, positioning it as a valuable precursor to AI-driven RTL debugging systems that aim to automate both bug localization and correction.

The **Silicon Debug Automation using Transaction Tagging and Data Mining** methodology introduces a scalable approach to post-silicon validation by bridging the gap between low-level signal traces and high-level design intent. It uses *transaction tagging*, where transactions in the design are marked with unique identifiers that propagate through the logic. These tags allow the debugger to track logical flows across the chip during execution. When errors are observed in silicon, the corresponding tags can be extracted from trace buffers and linked to their source or propagation path. By leveraging *data mining* techniques, such as pattern recognition and clustering, the system can analyze large volumes of trace data to detect anomalous or rare behaviors indicative of bugs. This approach significantly reduces manual analysis time and enables a semi-automated or fully automated identification of root causes in complex SoC designs [5].

RTL debugging methods can be broadly categorized into **rule-based**, **machine learning (ML)-based**, and **large language model (LLM)-based** approaches. Each represents a different philosophy in how verification and debugging are conducted, ranging from deterministic logic to data-driven pattern recognition to reasoning via natural language understanding (Fig. 4.1) [6].

4.2 Rule-based RTL Debugging

Rule-based RTL debugging is the traditional and most widely used approach. It relies on manually written assertions, formal properties, and static analysis rules to detect bugs and design violations. Tools like SystemVerilog Assertions (SVA), formal verification platforms, and linting tools fall under this category. These methods offer high precision and

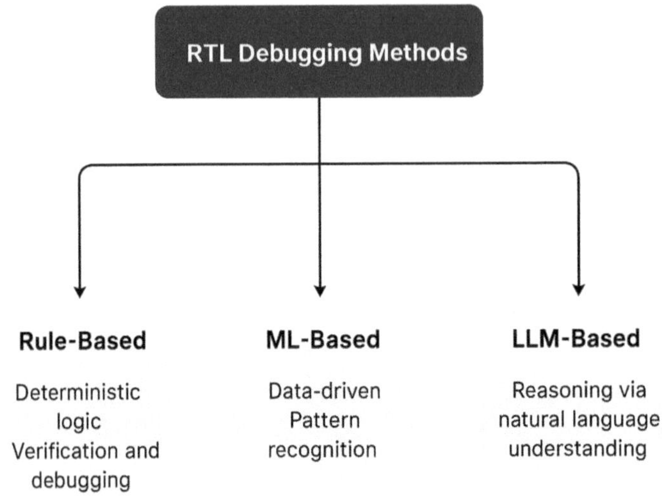

Fig. 4.1 RTL debugging methods

are well understood by engineers, but they tend to be rigid and require significant manual effort. They also struggle with detecting bugs that were not anticipated by the defined rules, making them less adaptable in large or evolving designs. We discussed this thoughtfully in Chap. 3.

4.3 Machine Learning: An Introduction

Types of Machine Learning are: Supervised learning, Unsupervised learning, Semi-supervised learning, and Reinforcement/**additive** learning (Figs. 4.2 and 4.3). In the supervised ML, a machine learns from past data and then produces the desired output. The supervised learning algorithms can be classified into two, namely regression and classification. Regression is a supervised learning technique whose output variables are continuous and can be assigned any real value that is regression type. The examples are weight, size, and amount. Classification is a type of supervised learning in which the output vectors are usually discrete or non-continuous such as gender and color. There are three kinds of classification task. Binary classification where there are only two exclusive classes. Multiclass classification where there are more than two classes, and the classifier can only report one of them as output. Multilabel classification where a combination of multiple independent binary classifiers [7–9].

When a machine learns from unlabeled data or it discovers the input pattern itself, it is known as unsupervised learning. When the machine learns from both labeled and unlabeled data, it is known as semi-supervised learning. Researchers have found that unlabeled data, when used in conjunction with a small amount of labeled data, can produce a

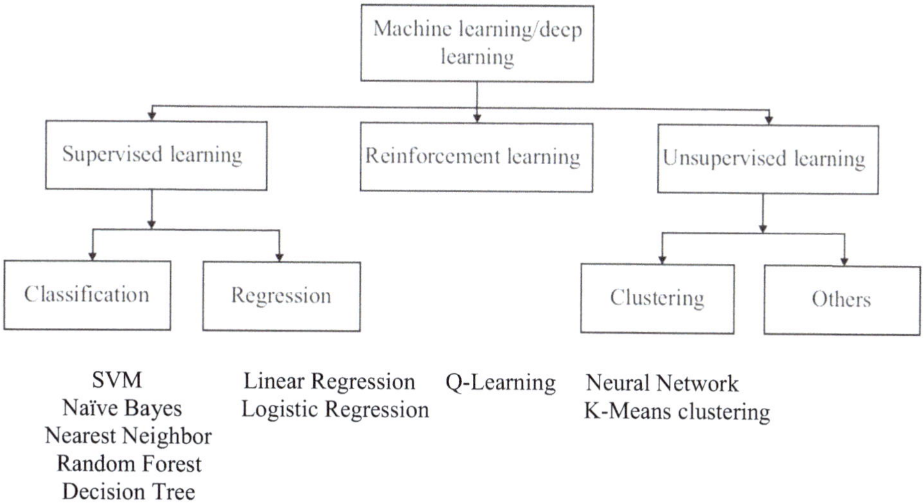

Fig. 4.2 Machine learning types

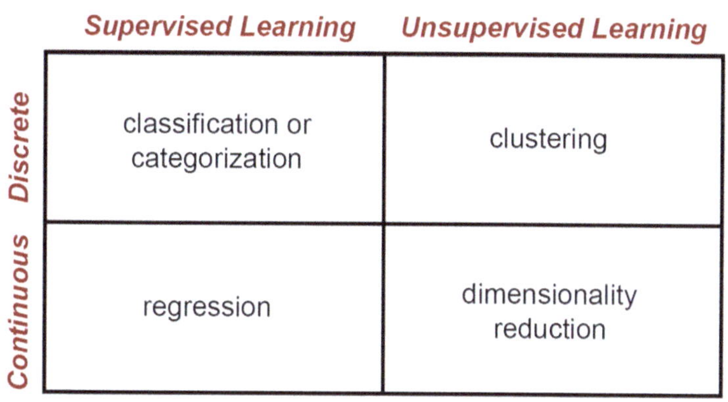

Fig. 4.3 Machine learning types vs data types

considerable improvement in learning accuracy. Reinforcement learning does not need training examples. In the reinforcement learning, models are given an environment, group of some actions, a goal, and a reward. This algorithm is learned by rewards and penalties. ML applications are summarized in Table 4.1.

Computer Vision (CV) and Pattern Recognition is one of the earliest applications of DL. Imageclassification, object detection, object/instance segmentation, image reconstruction, and captioning are the scopes of CV models. Natural Language Processing (NLP) deals with processing sequential data. Recurrent Neural Networks (RNN), Long-Short-Term-Memory (LSTM) and Gated Recurrent Unit (GRU) have been used in language modeling until the state-of-the-art Transformer model is introduced. NLP models are used in machine translation, text summarization, speech recognition, syntactic and

Table 4.1 ML applications

Machine learning type	Application
Supervised learning	• *Face Recognition* • *Pattern Recognition* • *Speech Recognition* • *Email Spam Filtering* • *Handwriting Recognition* • *Intrusion Detection*
Unsupervised learning	• *Social Network Analysis* • *Analysis of Cancer Diagnosis*
Semi-supervised learning	• *Sentiment Analysis*
Reinforcement learning	• *Computer Games* • *Stock Market Analysis* • *Robotics*

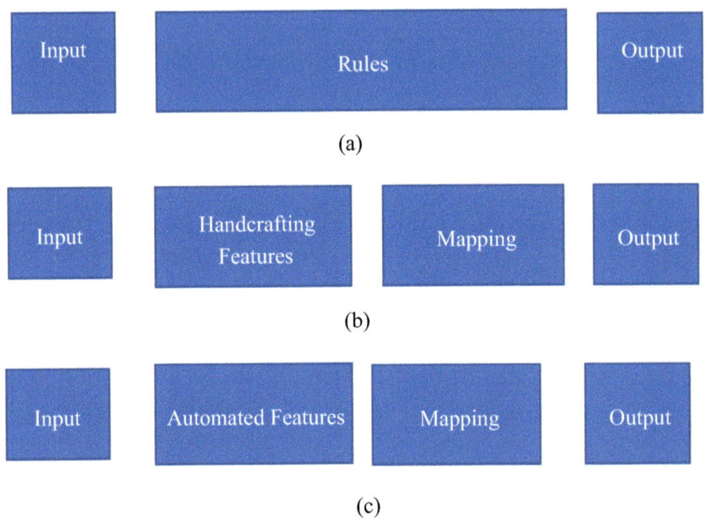

Fig. 4.4 (**a**) Rule-based learning, (**b**) feature-based learning, (**c**) and representation learning

semantic parsing, question answering, dialog system etc. There are three classifications of machine learning: rule-based learning, feature-based learning, and representation learning (Fig. 4.4).

Spark NLP is an open-source natural language processing (NLP) library built on top of Apache Spark, a popular big data processing framework. It provides a high-level API for building scalable and efficient NLP pipelines on large datasets, leveraging the distributed processing capabilities of Spark. Spark NLP offers a wide range of NLP functionalities, including tokenization, part-of-speech tagging, named entity recognition, sentiment analysis, and text classification. It also supports pre-trained models for several NLP tasks, such as word embeddings and language detection. Spark NLP is designed to work with several programming languages, including Python, Scala, and Java. It can be used with several big data platforms, including Apache Hadoop, Amazon EMR, and Google Cloud Dataproc. One of the key benefits of Spark NLP is its scalability. It can process large volumes of text data in a distributed and parallelized manner, making it suitable for handling big data NLP tasks. It also provides a unified API for NLP tasks, making it easier to build and maintain NLP pipelines.

Classical ML techniques can make use of the data volume to some magnitude. Their performance will not improve further even if we add more data. DL methods can use big data by building a complex model to learn nonlinear relationships between data. Analogy between human brain and different types of learning is shown in Fig. 4.5.

In the context of machine learning, the level of automation for **batch, real-time, and near real-time processing** is typically related to the speed at which data is processed and analyzed by machine learning models. Batch processing in machine learning involves processing a large dataset in batches or chunks, often during off-peak hours or in the background. This type of processing is commonly used for tasks such as training machine learning models or running analytics on historical data. Batch processing can be highly automated using tools such as Apache Spark, which can distribute and parallelize computations across a cluster of machines. Real-time processing in machine learning involves processing data as it arrives in real-time, with minimal delay. This type of processing is

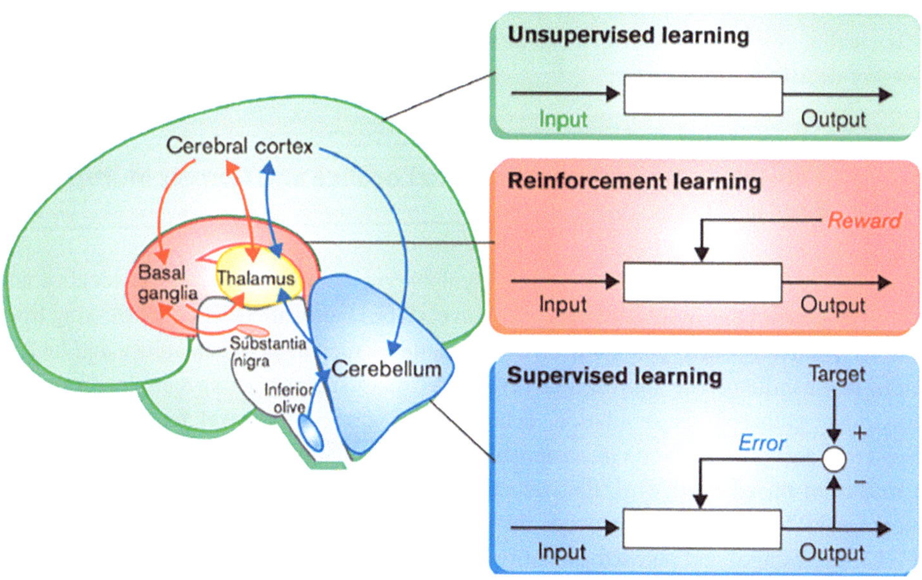

Fig. 4.5 Analogy between human brain and different types of learning. In the human brain, a typical neuron collects signals from others through a host of fine structures called *dendrites*. The neuron sends out spikes of electrical activity through a long, thin stand known as an *axon*, which splits into thousands of branches. At the end of each branch, a structure called a *synapse* converts the activity from the axon into electrical effects that inhibit or excite activity in the connected neurons

commonly used for applications such as fraud detection or anomaly detection, where immediate action is required based on incoming data. Real-time processing requires highly automated systems that can rapidly process and analyze incoming data, such as stream processing frameworks like Apache Flink or Apache Kafka. Near real-time processing in machine learning involves processing data in smaller batches, typically every few minutes, to provide faster results than traditional batch processing but not as fast as real-time processing. This type of processing is commonly used in applications such as predictive maintenance or recommendation systems. Near real-time processing can be automated using tools such as Apache Beam or Apache NiFi, which can process data in near real-time by leveraging batch processing techniques.

4.4 ML-Driven RTL Debugging

ML-based RTL debugging introduces data-driven intelligence to the verification process. These approaches use supervised or unsupervised machine learning algorithms trained on RTL simulation data, signal traces, or historical bug reports to predict or localize potential design errors. For example, classifiers can be trained to identify buggy modules based on signal behavior, or anomaly detection models can flag unexpected waveform patterns. ML-based debugging can scale effectively and uncover hidden patterns, but it depends heavily on the quality and quantity of training data. It is also less interpretable, often acting as a black box, which can be a challenge for engineers needing clear justification for decisions.

4.4.1 A Mutation-Based Framework to Localize and Correct Multiple Bugs in RTL

The work in [10] is a fully automated and scalable methodology designed to localize and correct multiple bugs in Register-Transfer Level (RTL) hardware designs, addressing limitations in existing debugging and correction methods such as poor scalability and lack of multi-bug handling. The approach involves three phases: constructing samples from coverage data gathered via simulation, debugging by applying a novel feature burst (FB) clustering algorithm to group hit-statements likely to contain bugs, and correction through a **mutation-based** framework that selectively fixes the highest-ranked suspicious statements. FixRTL significantly reduces the amount of code developers need to inspect (by 44.3% on average) and outperforms prior works by correctly fixing 67% of injected bugs compared to up to 25% in recent methods. Additionally, it supports hierarchical RTL designs and provides corrections that closely match the actual faults, making it a promising tool for efficient multi-bug debugging and repair in complex hardware verification workflows.

4.4.2 GA-Based Framework to Localize and Correct Multiple Bugs in RTL

According to Moore's law, designs are growing in size, with the number of transistors doubling approximately every 1.5 years. Additionally, the increasing complexity is widening the designer's productivity gap, making design efforts more expensive. Verification remains the primary bottleneck in modern VLSI development due to escalating circuit complexity.

Therefore, improving verification efficiency in terms of time, quality, and cost has become essential. As designs grow more complex, verification faces greater challenges. There are primarily two levels: IP-level verification, focused on functionality, and SoC-level verification, which emphasizes connectivity [11, 12].

Verification cycles are tightly constrained by time-to-market pressures. Within the IP life cycle, debugging involves locating and correcting design errors: a process that can consume nearly half of the development timeline. Traditional debugging relies on waveforms, reference outputs, and assertions, making it time-intensive.

Automated debugging is a promising solution to shorten development cycles and bridge the productivity gap. Bugs may reside in the design or in the verification environment. Verification techniques include both dynamic (simulation) and formal (property/equivalence) methods. Yet, existing EDA tools lack effective automatic correction mechanisms. This paper introduces an automated method for correcting RTL errors using mutation techniques.

Mutation testing, originally proposed in [13] for software quality assessment, injects faults to measure test effectiveness. In hardware, it has been applied at the gate level for error correction in prior works [11, 12]. However, its computational demands: due to separate simulations for each mutant—pose a major challenge. In this work, for the first time, mutation is applied at the RTL level for correction.

In the proposed automated debug flow [14], an RTL error is first detected and localized, after which a correction step is applied. The complete process integrates four core technologies, as summarized in Fig. 4.6. Once a candidate fix is applied, the corrected design must be re-verified to ensure correctness [15, 16]. This process continues iteratively until the design passes verification or all correction options have been exhausted.

If the design remains faulty after attempting all available mutants, the **FixRTL** tool reports that it is unable to correct the bug. This fail-safe ensures the user is informed of unresolvable issues under the current mutation model.

The operational procedure of the **FixRTL** tool is described in Listing 4.1, with an overview illustrated in Fig. 4.7. The methodology supports correction of multiple concurrent errors. The correction mechanism is based on mutation; wherein faulty statements are replaced with alternative versions generated by applying mutation operators from a predefined library.

A mutant is created by introducing small, syntactic changes to the RTL source code. Mutation operators are categorized based on the language construct they modify. For

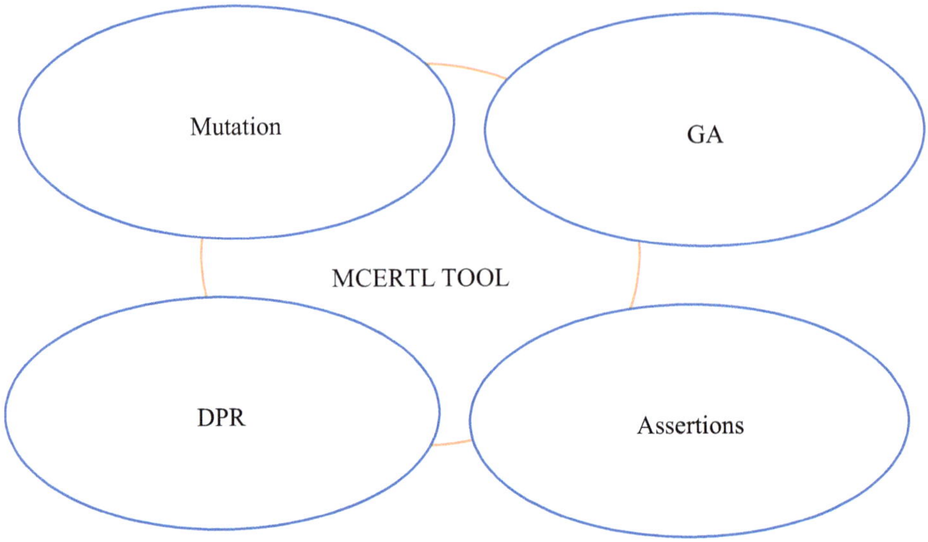

Fig. 4.6 The four technologies used in this work

example, arithmetic operators include addition, subtraction, multiplication, division, and modulo. If an arithmetic error is suspected, the faulty operator is systematically replaced with all other candidates in the arithmetic group.

In addition to arithmetic mutations, logical mutations (AND, OR, XOR, NOT, NAND) are supported, as well as relational operators such as equal (==), not equal (!=), greater than, less than, and their inclusive forms. Control flow constructs and constant values can also be mutated. A full list of supported mutation operators is provided in Table 4.2.

To illustrate the methodology, Table 4.3 presents an example of an erroneous up-counter that behaves like a down-counter. By applying concurrent arithmetic mutations, as shown in Fig. 4.8, the bug is identified and automatically corrected with minimal effort.

The proposed mutation-based correction technique is implemented as an emulation-based approach. Numerous experiments have been conducted to evaluate the effectiveness of this methodology, with the results summarized in Table 4.3. In the experimental setup, we consider three RTL code instances: the faulty RTL code (RF), the desired correct RTL code (RC), and a mutant version of the RTL code (RM). A correction is deemed successful when RM = RC. If RM ≠ RC, that mutant is discarded from the list of potential fixes. The selection of the mutant operator's type, class, or family plays a crucial role in determining the efficiency of the correction process. In this study, we focus on scenarios involving a single error spread across multiple lines within the RTL code. The effectiveness of the approach under different error counts is presented in Table 4.4. Although the correction run-time can be significant, it is negligible compared to the manual debugging time required by a human engineer. Additionally, the hardware overhead introduced by FixRTL is minimal: the area consumed by the correction logic is <2% of the total FPGA fabric. For reference, the original design itself occupies approximately 12% of the FPGA area, as shown in Table 4.5, based on an eMMC protocol implementation [11, 13, 15–29].

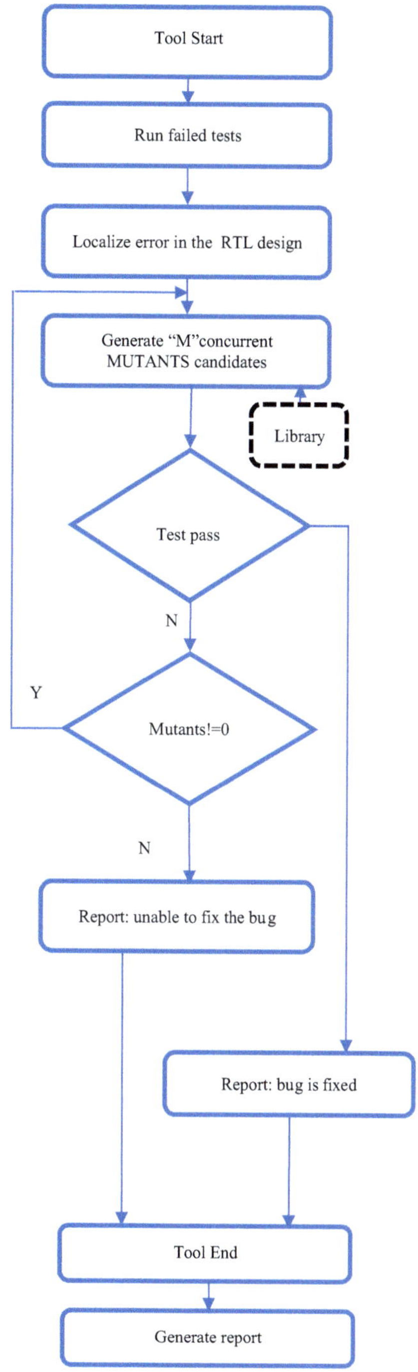

Fig. 4.7 MCERTL tool operational procedure

Table 4.2 Mutants class

Mutants class	Example
Arithmatic	+, −, *, /
Logical	‖, &&, ~
Relational	<, <=, >=, ==, !=
Bitwise	˜, &, |, ∧
Constants in loops	[0:255]

Table 4.3 Example to illustrate the mutation-based correction methodology

The code status	Test status

Input code with an error in the blue line
module up_counter (out, enable, clk, reset);
output [7:0] out;
input enable, clk, reset;
reg [7:0] out;
always @(posedge clk)
if (reset) begin
out <= 8'b0;
end else if (enable) begin
out <= out - 1;
end
	Fail

The code after applying Mutant #1.
module up_counter (out, enable, clk, reset);
output [7:0] out;
input enable, clk, reset;
reg [7:0] out;
always @(posedge clk)
if (reset) begin
out <= 8'b0;
end else if (enable) begin
out <= out * 1;
end
	Fail

The code after applying Mutant #2: The code is fixed
module up_counter (out, enable, clk, reset);
output [7:0] out;
input enable, clk, reset;
reg [7:0] out;
always @(posedge clk)
if (reset) begin
out <= 8'b0;
end else if (enable) begin
out <= out + 1;
end
	Pass

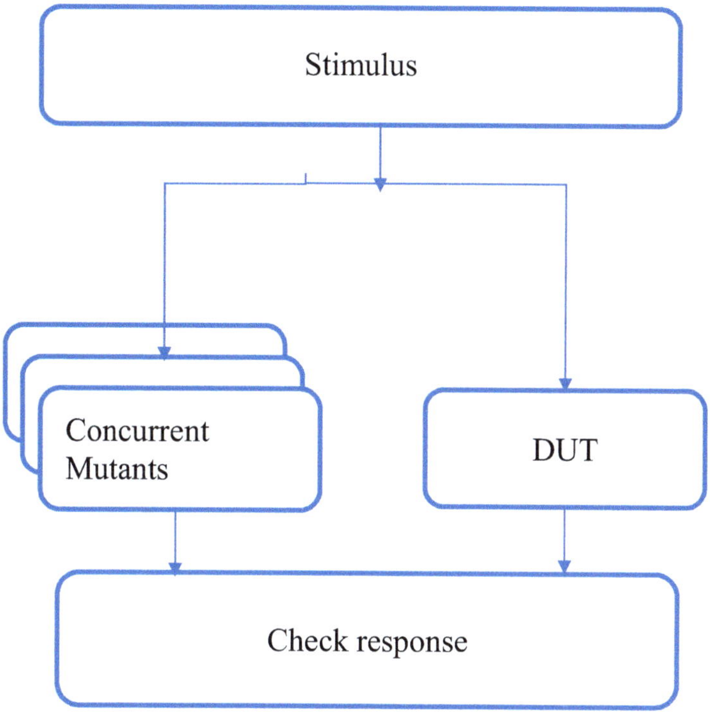

Fig. 4.8 Concurrent mutants correction methodology

4.5 Large Language Model (LLM): An Introduction

Listing 4.1 Pseudo-Code for Operational Procedure

```
localize potential bugs;
   Generate all possible Mutations;
   FOR (all mutants)
   run parallel tests
  IF (test is passing)
   Exit;
   ELSE "BUG cannot be corrected";
   END IF
   END FOR
```

A large language model (LLM) is a type of machine learning model that can perform a variety of natural language processing (NLP) tasks, including generating and classifying text, answering questions in a conversational manner and translating text from one language to another (Fig. 4.9). The label "large" refers to the number of values (parameters)

Table 4.4 Error correction experiments: the errors are artificially generated

Design	# errors	Processing time (s)	# Examined mutants
Up-Counter	One	100	17
CRC		200	24
GCD		259	36
EMMC		400	37
SDCard		370	35
Up-Counter	Two	220	22
CRC		310	34
GCD		457	40
EMMC		612	46
SDCard		600	40
Up-Counter	Three	370	25
CRC		490	32
GCD		589	47
EMMC		656	43
SDCard		600	40
Up-Counter	Four	300	31
CRC		300	38
GCD		459	42
EMMC		500	54
SDCard		467	49
Up-Counter	Five	500	42
CRC		600	56
GCD		659	57
EMMC		700	66
SDCard		689	61
Up-Counter	Six	600	59
CRC		657	52
GCD		722	57
EMMC		814	66
		723	59
Up-Counter	Seven	653	66
CRC		769	73
GCD		812	75
EMMC		908	83
SDCard		899	81

the model can change autonomously as it learns. Some of the most successful LLMs have hundreds of billions of parameters. Large Language models use reinforcement learning for fine-tuning. Reinforcement Learning is a feedback-driven Machine learning method based on a reward system. An agent learns to perform in an environment by completing certain tasks and observing the results of those actions. The agent gets positive feedback for every good task and a penalty for each bad action. However, the combination of the supervised learning and reinforcement learning can attain optimal performance. Examples of LLMs include Generative Pre-trained Transformer 3 (GPT-3), Bidirectional Encoder

Table 4.5 Synthesis results for emmc protocol with/without assertions on XILINX ARTIX-7

Utility	Without assertions	With assertions	FPGA (DUT)	FPGA (DUT+SVA)
Slice LUT	55	74	0.04%	0.05%
Slice registers	72	87	0.02%	0.03%
Flip flop registers	58	67	0.02%	0.02%
Latch registers	14	20	<0.01%	0.01%
I/O ports	41	47	10.25%	11.75%

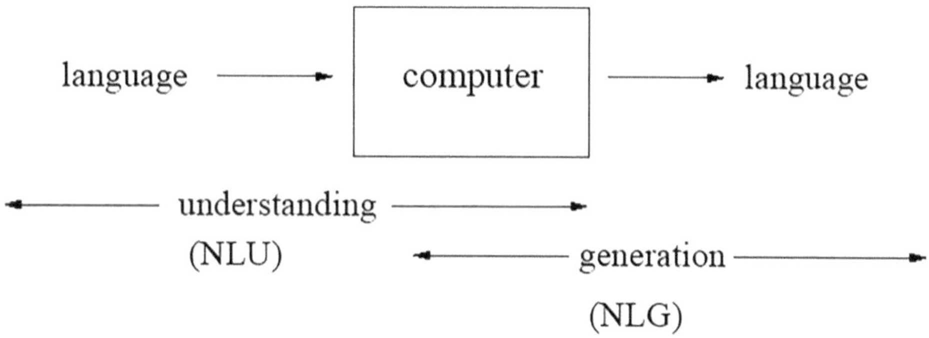

Fig. 4.9 NLP representation: NLU, NLG

Representations from Transformers (BERT), Large Language Model Meta AI (LLaMA), and multimodal large language model (MLLM) [30–34].

sIn the context of deep learning, a **tokenizer** is a crucial component for processing textual data, which is commonly used in natural language processing (NLP) tasks. A tokenizer is a function or a module that breaks down a string of text into individual tokens or words, usually by separating them based on whitespace or punctuation marks. This process of tokenization converts raw text into a sequence of numerical values that can be easily fed into a neural network for training and inference. Tokenization is an essential step in many NLP tasks, including sentiment analysis, text classification, and language translation. Deep learning models rely heavily on tokenization to preprocess textual data and convert them into a format that can be effectively processed by the neural network. There are various types of tokenizers used in deep learning, including word-level tokenizers, character-level tokenizers, subword-level tokenizers, and byte-pair encoding (BPE) tokenizers. Each type of tokenizer has its strengths and weaknesses, depending on the specific task and the characteristics of the data being processed. **LLM accelerates debugging cycle** [35].

4.5.1 GPT

OpenAI introduced Generative Pretrained Transformer (**GPT**). It is a very large language model trained on a massive dataset from the internet and Wikipedia. It was pretrained on a lot of text from the Internet and thus ready to respond to questions, it uses a particular kind of neural network architecture called the **Transformer** [92–94]. The transformer neural network is a novel architecture that aims to solve sequence-to-sequence tasks while handling long-range dependencies with ease (Fig. 4.10). It was proposed to overcome the disadvantages of RNN such as slow training time and vanishing gradient. Transformers process the entire input all at once unlike RNNs. It is major application is NLP. **GPT-3** was first introduced in June 2020. It is trained on a massive dataset of over 570GB of text data, which includes a diverse range of sources such as books, articles, and websites. GPT-3 has 175 billion parameters, making it the largest language model to date. GPT-4 is able to process multiple types of data including Videos, Images, Sounds, Numbers etc.

4.5.2 BERT

Google introduced Bidirectional Encoder Representations from Transformers (BERT). It makes use of Transformer, an **attention mechanism** that learns contextual relations between words in a text [78]. In its vanilla form, Transformer includes two separate mechanisms: an encoder that reads the text input and a decoder that produces a prediction for the task. The difference between GPT-3 and BERT is their architecture. GPT-3 is an

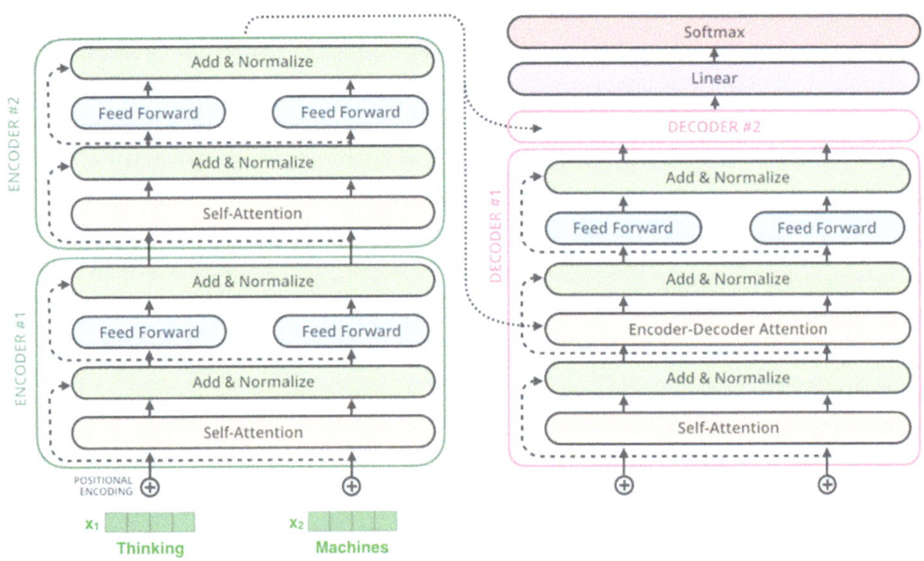

Fig. 4.10 Transformer architecture: stacked encoders and decoders. A transformer uses Encoder stack to model input and uses decoder stack to model output using input information from encoder side. If we do not have input, we just want to model the next word, we can get rid of the Encoder side of a transformer and output "next word" one by one. This gives us **GPT**. If we are only interested in training a language model for the input for some other tasks, then we do not need the Decoder of the transformer, that gives us **BERT**

autoregressive model, while BERT is bidirectional. GPT-3 considers only the left context when making predictions, BERT considers both left and right context (Figs. 4.11 and 4.12). BERT is pretrained on massive data and is used in General domain applications.

4.5.3 LLaMA

Meta introduced Large Language Model Meta AI (LLaMA). It is a collection of foundational large language models ranging from 7 to 65 billion parameters. LLaMA is creating excitement because it is smaller than GPT-3 but performs better. One of the most important features of LLaMA is that it is available under a non-commercial license to researchers and other organizations, making it easier for them to use it for their work.

4.5.4 MLLM

Microsoft has introduced Kosmos-1, a state-of-the-art multimodal large language model (MLLM) designed to process and understand multiple modalities, including both images and text. Unlike traditional models that typically handle a single modality, Kosmos-1 integrates information across these different input types to perform a wide range of tasks. It

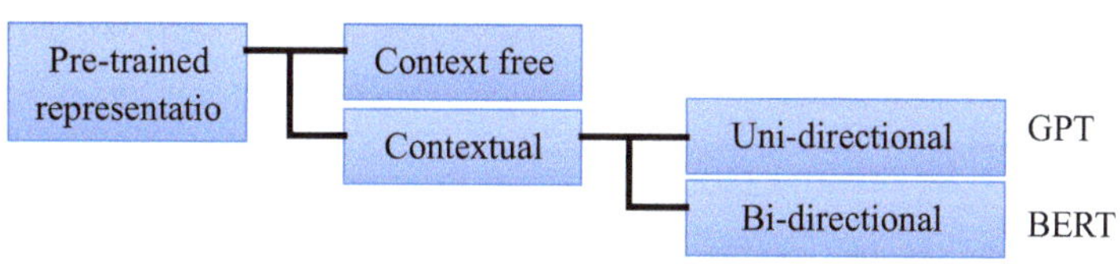

Fig. 4.11 GPT vs BERT

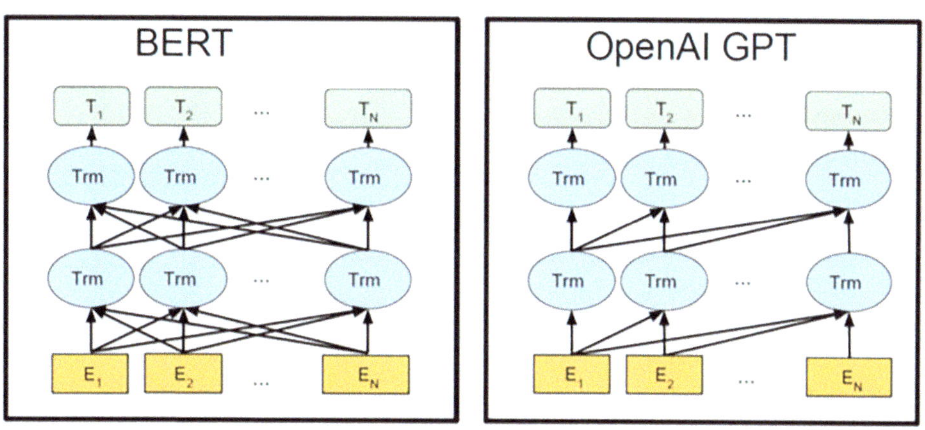

Fig. 4.12 GPT vs BERT. BERT uses a bidirectional transformer. GPT uses a left to right transformer

can answer questions by interpreting mixed inputs, effectively combining visual and textual data to provide contextually relevant responses. Beyond simple question answering, Kosmos-1 demonstrates advanced visual dialogue capabilities, enabling interactive conversations grounded in visual content. This enables the model to maintain context and coherence across multiple turns of a dialogue involving images. Furthermore, Kosmos-1 exhibits promising performance in more sophisticated reasoning tasks such as visual IQ tests, which require understanding complex visual scenes and applying logical inference. These advancements illustrate the growing capability of multimodal models to bridge the gap between vision and language, pushing forward the frontiers of artificial intelligence in understanding and reasoning over diverse data formats [34].

4.6 LLM-Driven RTL-Level Debugging

4.6.1 Introduction

The rapid advancements in LLMs have driven the adoption of generative AI in various domains, including Electronic Design Automation (EDA) [36–38]. In recent years, large language models (LLMs) have demonstrated remarkable performance in natural language processing (NLP). Inspired by this progress, researchers have started exploring the adoption of LLMs in agile hardware design [39, 40].

LLM-based RTL debugging represents a more recent and powerful paradigm, leveraging large language models like GPT or Claude to understand RTL code, simulation logs, waveform data, and even design specifications in natural language. These models can analyze state checkpoint logs, identify mismatches between expected and actual behavior, and suggest code-level fixes. Because they can reason across diverse inputs and provide explanations in human language, LLMs are especially useful for complex debugging tasks and collaborative verification. However, they can sometimes produce incorrect or inconsistent responses if not guided properly and typically require structured prompts or logs to be effective [24].

Despite the strong capabilities of Large Language Models (LLMs) in generating, completing, and debugging software code, their utilization in the specialized field of HDL debugging has been limited and, to date, has not yielded satisfactory results [41]. Unlike software languages like Python or C++, where code can often be crawled from open websites and platforms like GitHub, HDL codes, particularly those used for chip testing, are rarely made public due to privacy and commercial concerns.

Figure 4.13 illustrates a high-level abstraction of a "Debug Core" system designed to automate and streamline the RTL debugging process. It accepts four distinct inputs: the hardware specifications (Specs), buggy RTL code, simulation logs, and compilation logs. These inputs represent the core elements of a typical hardware design debugging workflow. The specifications guide the intended functionality, while the buggy RTL code forms the initial, flawed implementation of the design. Simulation and compilation logs provide

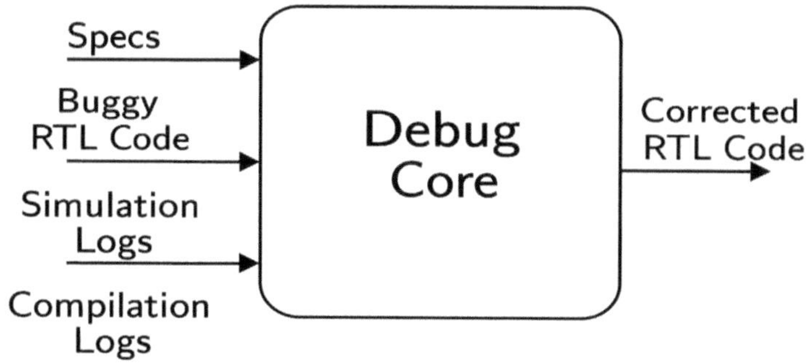

Fig. 4.13 A generic framework for LLM-based RTL debug core

critical feedback by capturing functional mismatches, syntax issues, timing errors, and other anomalies encountered during verification and synthesis. The Debug Core, depicted centrally in the diagram, processes these inputs to analyze the nature and source of bugs. Through contextual interpretation and cross-referencing of design intent (Specs) with observed failures (Logs), the core applies corrective actions to the RTL code. The result is a new, error-resolved version of the RTL design, output as "Corrected RTL Code." This framework reflects a shift toward intelligent, LLM-assisted or tool chain-enhanced debugging strategies, where automation significantly reduces manual iterations, improves turnaround time, and enhances productivity in hardware development cycles [42].

4.6.2 AI-Driven RTL Debugging with Verification In-The-Loop: Iterative Refinement, Embedded Retrieval and Guided Bug Fixing Generation

The work in [43] designs a fully LLM-based RTL debugging framework that does not rely on any third-party closed-source tools, leveraging a state checkpoint mechanism to improve diagnosis and correction of logic bugs in RTL designs (e.g., Verilog). Debugging is a critical part of RTL design and verification, and recent advances in large language models (LLMs) have opened new possibilities for automating and accelerating this process. Traditional RTL debugging typically involves meticulous code-level inspection using waveform analysis, assertions, and simulation traces. While effective for low-level signal issues, these methods often fall short when designers must reason about complex control logic, protocol-level behavior, or timing-sensitive corner cases. Early LLM-based efforts focused on line-by-line code understanding, which is helpful for syntax-level issues (e.g., missing signal declarations or incorrect assignments) but not sufficient for deeper bugs rooted in algorithmic intent or protocol violations. This has led to growing interest in **leveraging natural language reasoning** within RTL debugging workflows. By combining knowledge of digital design semantics with the ability to interpret high-level intent described in natural language, LLMs can potentially assist in diagnosing subtle bugs,

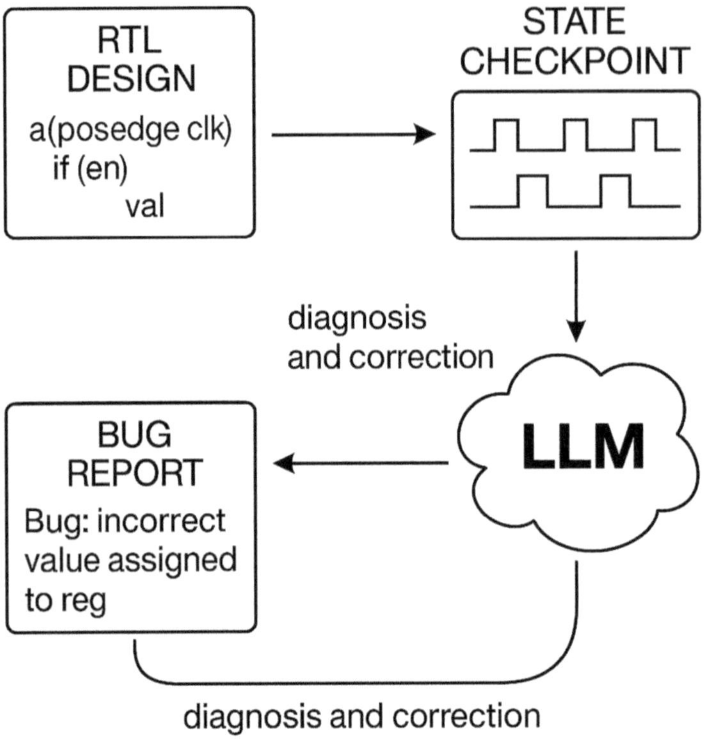

Fig. 4.14 Overview of methodology in [43]

guiding waveform inspection, or even suggesting logic fixes based on design specs and observed behavior (Fig. 4.14).

The work in [44] introduces MEIC (Make Each Iteration Count), a novel framework that leverages Large Language Models (LLMs) to automate RTL (Register Transfer Level) code debugging through an iterative, human-inspired process. Unlike traditional one-shot LLM approaches, MEIC incorporates a feedback loop using two specialized LLM agents—one for debugging and another for scoring—alongside a toolchain for compilation and simulation. This setup allows MEIC to progressively refine RTL code by correcting syntax and function errors while managing uncertainties in LLM behavior. Experimental results show that MEIC achieves 93% fix rate for syntax errors and 78% for function errors, outperforming previous LLM-based methods and even achieving up to 48× debugging speedup compared to human engineers. The framework also includes an open-source dataset and tooling, demonstrating significant promise for practical adoption in hardware verification and design workflows (Fig. 4.15).

In RTL simulations, a state checkpoint captures snapshots of internal signal values and outputs at various points in time during simulation. These checkpoints act as reference states, helping the LLM reason about what the design should be doing vs. what it actually did.

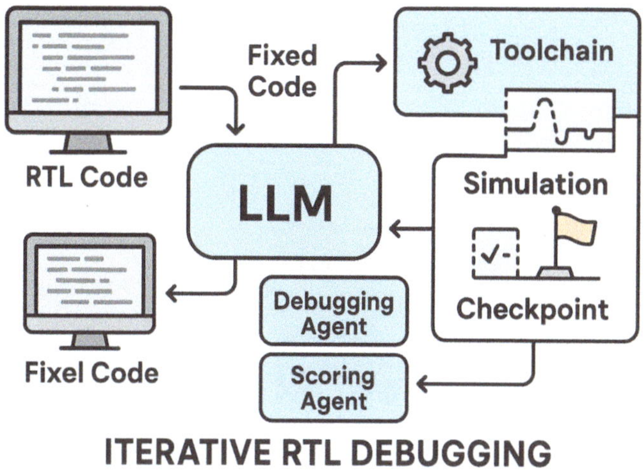

Fig. 4.15 Overview of methodology in [44]

The work in [45] introduces a unified large language model (LLM) framework specifically designed to automate Verilog RTL debugging by combining contrastive embedding and guided generation techniques. Unlike traditional generative-only approaches prone to hallucinations, VeriDebug first encodes code and specifications into embeddings to retrieve relevant, semantically similar internal information, then uses this information to generate accurate bug fixes. This joint approach enables it to simultaneously learn patterns of bugs and their corrections within a shared model space, significantly improving both bug localization and fixing generation. By avoiding dependence on external knowledge bases and leveraging open-source models, **VeriDebug** ensures secure, transparent, and context-aware debugging—demonstrating a major leap in the application of LLMs to hardware design, achieving a bug fix accuracy of 64.7%, far surpassing both open- and closed-source baselines (Fig. 4.16). LLMs are capable of analyzing Verilog code to identify syntax errors, logical inconsistencies, and potential design issues, while also providing suggestions for improvement [46, 47].

The work in [48] proposes a novel framework that leverages Large Language Models (LLMs) to assist in debugging Verilog RTL code by enabling *self-verification* and *self-correction*. In the context of RTL debugging, the core idea is to use LLMs not just as passive code generators but as intelligent agents capable of understanding, verifying, and correcting RTL code autonomously. The system can detect mismatches between RTL and intended behavior described in natural language or testbenches and iteratively refine the RTL to align with expected functionality. This approach significantly reduces manual debugging efforts, accelerates development cycles, and highlights the growing role of AI in automating complex hardware design tasks (Fig. 4.17).

The work in [49] introduces a framework that integrates AI language models with formal and simulation-based verification to generate and iteratively refine Verilog RTL code.

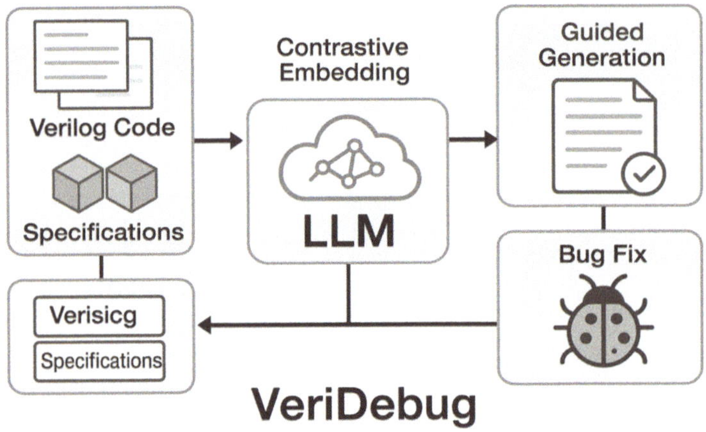

Fig. 4.16 Overview of VeriDebug

Fig. 4.17 Overview of methodology in [48]

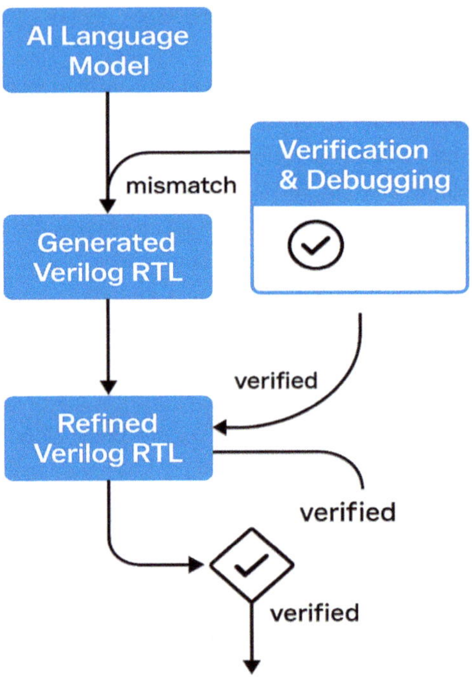

In the context of RTL debugging, the key idea is that instead of relying solely on human-written RTL, it places verification directly in the generation loop; automatically checking correctness and guiding the model to fix errors in the generated code. This tight feedback cycle between code generation and verification enables early bug detection, reduces manual debugging overhead, and ensures that the synthesized RTL not only matches the design intent but also passes formal and dynamic correctness checks (Fig. 4.18).

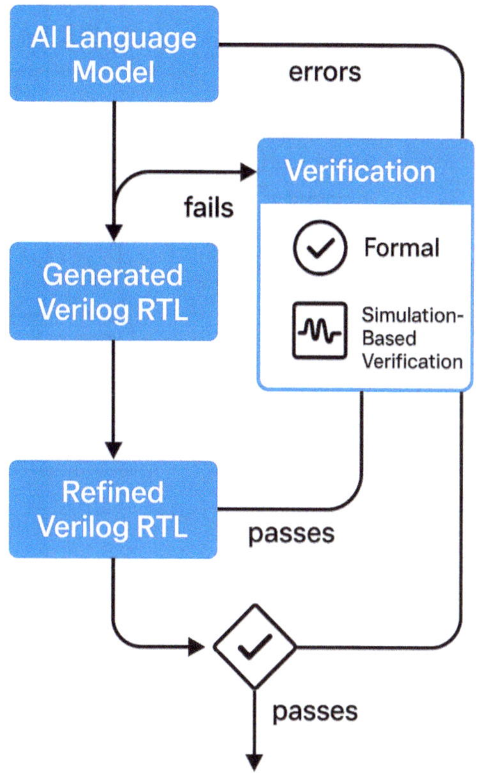

Fig. 4.18 Overview of methodology in [49]

The work in [50] introduces **VeriAssist**, an LLM-driven framework designed to autonomously generate, verify, and iteratively refine Verilog RTL code. Unlike traditional one-shot LLM code generation approaches that often fail to meet hardware timing and functionality requirements, VeriAssist mimics the step-by-step methodology of a human designer. It begins with structured prompts to understand design intent, then generates initial RTL code and testbenches, followed by *self-verification*; walking through code behavior over time steps—and *self-correction* based on real simulation feedback. Through iterative refinement using a simulation-in-the-loop process, the model adjusts its output to address both syntax and functional errors. This interactive loop enhances the model's ability to reason over hardware timing and logic, resulting in significantly higher correctness rates and reduced manual debugging overhead. Overall, VeriAssist advances RTL debugging by transforming LLMs from static code generators into adaptive design partners capable of contextual reasoning and improvement across iterations (Fig. 4.19).

The work in [51] presents an innovative approach to automatically generate functional testbenches for hardware description language (HDL) designs by leveraging large language models (LLMs) combined with a novel self-correction mechanism. The methodology begins by taking the RTL design specifications expressed in natural language and using an LLM to generate initial testbenches. Recognizing that LLM-generated

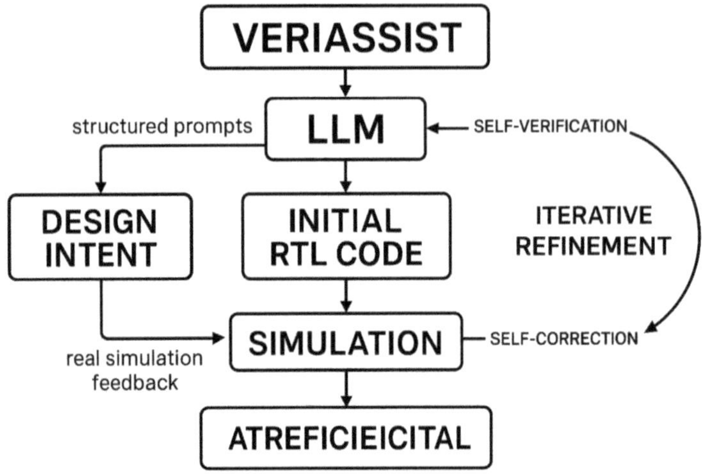

Fig. 4.19 Overview of VeriAssist

testbenches can contain functional errors, CorrectBench integrates a functional self-validation step that simulates the generated testbench against the RTL design to identify correctness issues or bugs. Instead of discarding faulty outputs, the system feeds the detected errors back into a specialized LLM-based corrector module, which uses the bug reports and failure information to iteratively refine and correct the testbench code. This closed-loop of generation, validation, and correction continues until the testbench passes functional validation or a maximum iteration limit is reached. By automating this feedback-driven correction process, the framework significantly improves the quality and functional accuracy of generated testbenches without requiring manually written test templates or extensive human intervention. The methodology is designed to work effectively even with complex sequential logic circuits, which are traditionally challenging for automatic testbench generation tools. This self-correcting pipeline thus enables more reliable, scalable, and automated verification workflows in HDL design environments (Fig. 4.20).

The work in [52] introduces a novel framework designed to enhance the generation of Register Transfer Level (RTL) code using Large Language Models (LLMs). While LLMs have shown promise in automating RTL code generation, challenges such as limited training data, misalignment between natural language specifications and generated code, and lack of verification mechanisms persist. To address these issues, **VeriReason** integrates supervised fine-tuning with Guided Reward Proximal Optimization (GRPO) reinforcement learning. This approach utilizes curated training examples and a feedback-driven reward model, combining testbench evaluations with structural heuristics to improve specification-code alignment and reduce errors. The iterative GRPO process embeds self-checking and reasoning capabilities, enabling the model to autonomously detect and correct functional errors. Experimental results on the VerilogEval Benchmark demonstrate significant improvements, with VeriReason achieving 83.1% functional correctness,

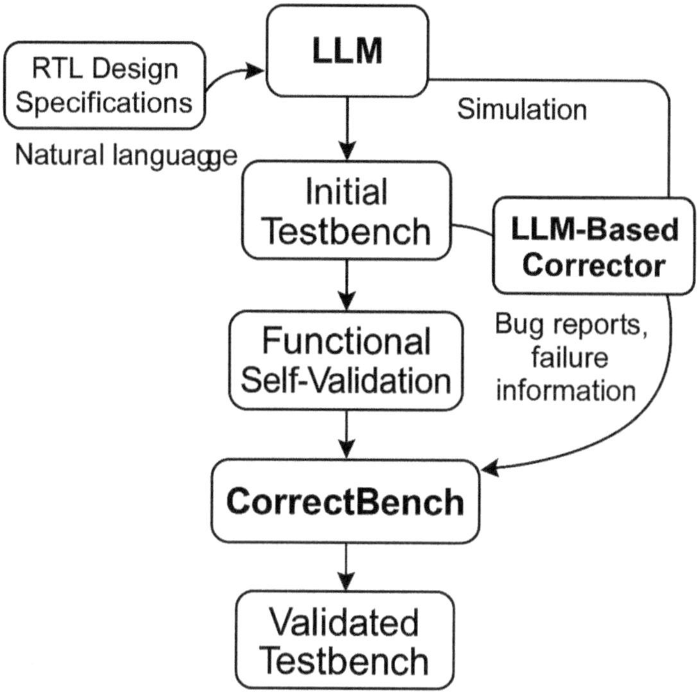

Fig. 4.20 Overview of methodology in [51]

outperforming both comparable-sized models and larger commercial systems like GPT-4 Turbo. Additionally, the framework shows up to a 2.8× increase in first-attempt functional correctness compared to baseline methods and exhibits robust generalization to unseen designs. This work represents a significant advancement in integrating reasoning capabilities with reinforcement learning for Verilog generation, setting a new state-of-the-art for automated RTL synthesis (Fig. 4.21).

The authors in [53] introduce **FLAG** (Finding Line Anomalies with Generative AI), a pioneering test-free and specification-free fault localization tool that leverages the power of **Large Language Models (LLMs)** to identify buggy lines in RTL code without requiring any runtime infrastructure. Motivated by prior findings that buggy code tends to be statistically "unnatural"—exhibiting higher entropy and deviating from the syntactic patterns of typical code—the authors extend this insight by proposing that LLMs, pre-trained on large volumes of open-source code, can highlight such anomalies through their token prediction behavior. FLAG performs both line-level and token-level analysis, comparing the original RTL code against LLM-generated alternatives and measuring the model's confidence in its predictions. If the model generates a token or line with high confidence that significantly differs from the original, that section is flagged as potentially defective. This approach builds on foundational research in software naturalness and entropy-based bug detection, while uniquely applying it to the hardware domain with the added

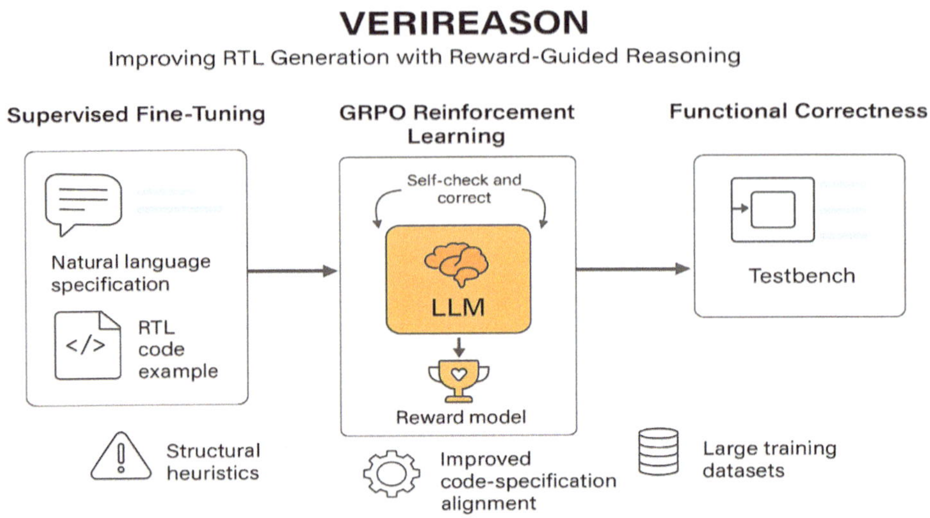

Fig. 4.21 Overview of VeriReason

advantage of not needing curated test sets or fault injection campaigns. Evaluated on a mix of real-world and synthetic benchmarks covering functional and security defects, FLAG demonstrates strong potential (Fig. 4.22).

The authors in [45] presents **VeriDebug**, a unified framework that leverages large language models for **automated Verilog debugging** by integrating **contrastive representation learning with guided correction mechanisms** (Fig. 4.23). Unlike prior LLM-based debugging tools that struggle with hallucination and overdependence on external resources, VeriDebug builds a **shared embedding space** that captures both bug patterns and their corresponding fixes, enabling the model to jointly perform **bug detection, localization, and code correction**. In the **representation phase**, Verilog inputs—including specifications and buggy code—are embedded into vector representations using contrastive learning to maximize similarity between true bug-fix pairs while distinguishing from negative examples. In the **generation phase**, the model retrieves relevant contextual information based on cosine similarity and uses it to generate corrected code, thereby reducing hallucination and improving accuracy. These two training objectives are combined in a joint loss function, empowering the model to learn bug identification and remedy in a unified manner.

4.6.3 LLM for Waveform Debugging

Recent advancements in AI-driven debugging have introduced innovative tools that leverage large language models (LLMs) to automate and enhance the waveform analysis process during hardware verification. These tools address the challenges of debugging

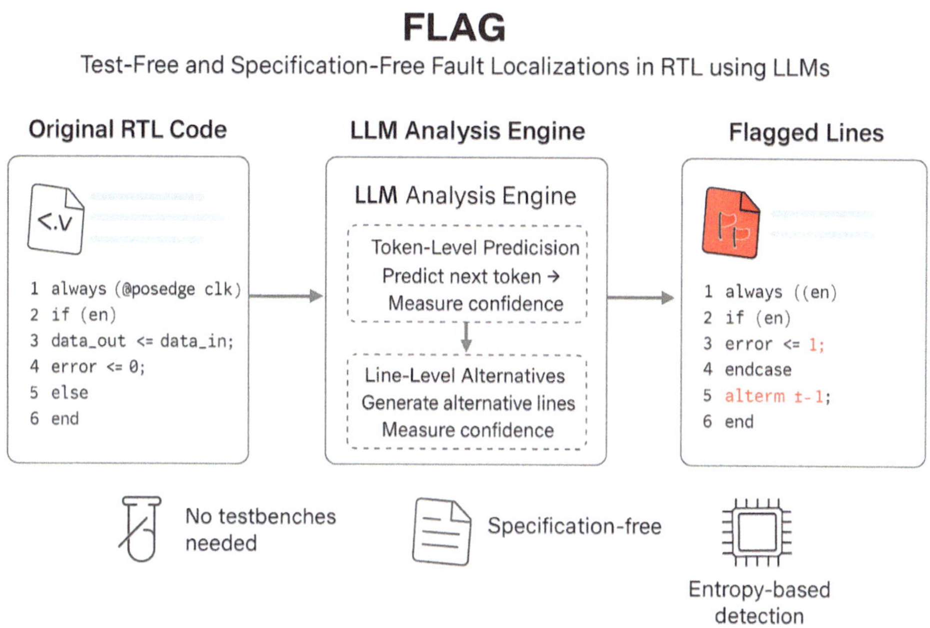

Fig. 4.22 Overview of FLAG

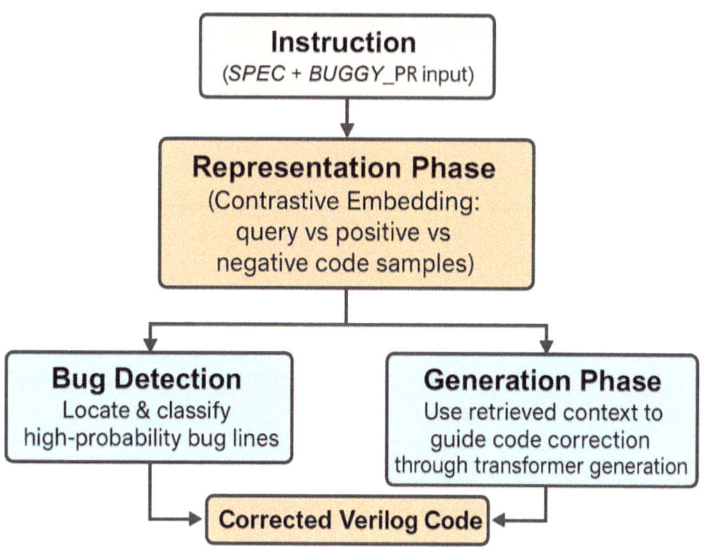

Fig. 4.23 Overview of VeriDebug

complex simulations by intelligently navigating waveform data, tracing signals, and cross-referencing design files. Unlike traditional debugging methods that require manual inspection of vast waveform data, this approach dynamically selects relevant contexts, enabling scalable and flexible analysis beyond the strict input size limits typically faced by standard LLMs. This methodology not only localizes faults but can also propose fixes, dramatically reducing the time engineers spend on waveform de-modeling. Benchmarking on realistic debugging tasks has shown that these AI-powered agents consistently outperform state-of-the-art LLMs in identifying and resolving bugs, with improved performance scaling as the reasoning budget increases. The integration of AI and hardware verification exemplifies how combining domain-specific knowledge with advanced machine learning techniques can revolutionize traditional workflows, making complex verification more efficient and accessible. The ability to query specific signals, verify invariants, and obtain precise timestamps further enhances usability, positioning such tools as essential components for continuous integration pipelines and automated bug triaging in modern chip development (Fig. 4.24).

4.6.4 LLM for Assertions Generation

SystemVerilog Assertions (SVAs) are vital tools for detecting and **debugging** functional bugs in digital chip design. However, the process of generating these assertions has traditionally been manual, time-consuming, and prone to errors. Recent advancements in automation, particularly those leveraging machine learning and large language models (LLMs); have shown promising potential to streamline assertion generation, though most existing methods remain in their early developmental stages. Authors in [54], present *Spec2Assertion*, a novel approach for automatically generating assertions directly from design specifications, even before the RTL is implemented. The proposed technique employs LLMs enhanced with progressive regularization and utilizes Chain-of-Thought (CoT) prompting to guide and improve the synthesis process. They also introduce a new evaluation methodology that measures assertion quality across diverse design scenarios. Experimental results on several benchmark designs demonstrate that Spec2Assertion produces 70% more syntax-correct assertions and achieves a twofold improvement in quality compared to a recent state-of-the-art method (Fig. 4.25).

4.7 Harnessing Natural Language in RTL Debug: Formats, Insights, and Limitation

As RTL designs grow in complexity, traditional debugging methods—heavily reliant on waveform inspection, manual tracing, and assertion failures; are becoming increasingly time-consuming and error-prone. Recent advancements in large language models (LLMs) have opened up new possibilities for streamlining the RTL debugging process by

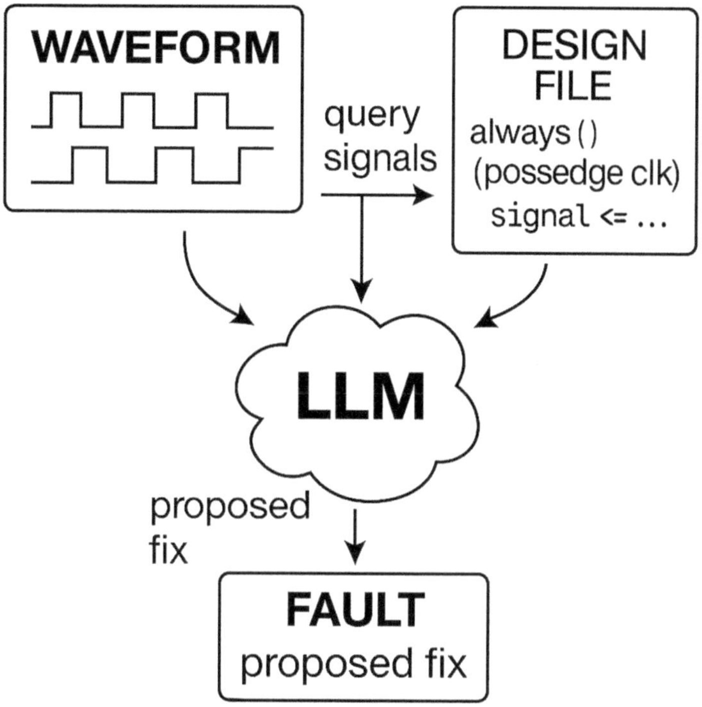

Fig. 4.24 Overview of LLM for waveform debugging

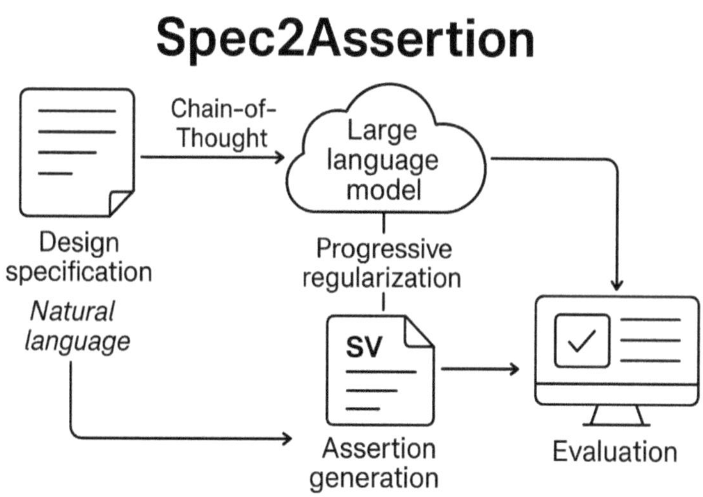

Fig. 4.25 Overview of LLM for assertions generation

integrating natural language interfaces into design workflows. These interfaces allow engineers to express design intent, describe failing behaviors, and query the system in a more intuitive, high-level manner. This section explores the effectiveness of various natural language formats in guiding RTL analysis, the specific benefits natural language reasoning brings to debugging, and the current limitations that constrain LLMs in practical hardware development. Collectively, these insights reveal both the potential and the boundaries of using LLMs as intelligent assistants for debugging RTL code.

4.7.1 What Type of Natural Language Format Is Most Effective for RTL Debugging Tasks?

Effective formats typically include structured descriptions of expected design behavior, failing scenarios, waveform observations, and formal specifications. For example, a natural language prompt such as "In the case of a read-after-write hazard, the read data must reflect the latest write data if no intermediate stall occurs" provides a clearer behavioral intent than signal dumps alone. When paired with simulation output (e.g., "data_out = 0×0 instead of 0xAB at cycle 112"), LLMs can map failure symptoms to likely RTL code regions or suggest causal signal chains. Formats that include assertions, transaction-level logs, or protocol descriptions also enrich the reasoning context for the model.

4.7.2 What Specific Benefits Does Natural Language Reasoning Bring to the RTL Debugging Process?

Natural language reasoning allows LLMs to bridge the gap between low-level RTL signals and high-level design intent. Instead of only responding to syntax or simulation mismatches, LLMs can infer protocol behavior, detect implicit violations (e.g., handshake mismatches, deadlock conditions), and even generate hypothesis-driven suggestions ("Did the designer forget to latch this value on the rising edge?"). This abstraction layer is especially valuable in debugging non-deterministic failures, timing-related bugs, or scenarios involving multiple interacting IP blocks. Furthermore, LLMs can help junior engineers ramp up quickly by explaining RTL constructs or identifying common design pitfalls based on the context of the bug.

In conclusion, integrating natural language reasoning into RTL debugging enriches the diagnostic process by making high-level design intent explicit, enabling deeper inference, and aligning code fixes with functional correctness. When paired with simulation outputs, waveform data, and partial specifications, LLMs can act as intelligent assistants—reducing debug time, improving design quality, and enhancing overall productivity in complex hardware development environments.

4.7.3 The Limitation of LLMs in RTL Code Debugging

Verification and debugging in hardware development are often twice as time-consuming as the design phase itself, highlighting the need for more efficient approaches. Large Language Models (LLMs) offer promising capabilities in this area, particularly in interpreting hardware specifications written in natural language and generating RTL code (e.g., Verilog, VHDL). While recent efforts have shown the potential of LLMs in automating hardware design, they also expose notable limitations. LLMs tend to perform more effectively in specialized domains when provided with system-level guidance and enriched with relevant domain knowledge [44, 55–58].

1. **Partial Understanding of Design Intent**
   LLMs may miss architectural nuances, timing constraints, and corner-case handling that are not explicit in code.
2. **Limited Awareness of Toolchain and Constraints**
   LLMs are unaware of synthesis/simulation tool behavior, timing constraints (e.g., SDC), or linting tool rules.
3. **Code Scope Limitation**
   Large RTL designs may exceed token limits, hindering the LLM‘s ability to trace signals across hierarchies.
4. **No Awareness of Functional Coverage or Assertions**
   LLMs cannot see functional coverage reports, assertion failures, or DUT-reference model mismatches unless described.
5. **Diagnosis Not Correction**
   LLMs suggest potential causes but cannot verify solutions against real hardware behavior.
6. **Dependent on User Guidance**
   Effectiveness relies on clear, detailed user input such as logs, waveforms, and interface specs.

References

1. Xu, K., Qiu, R., Zhao, Z., Zhang, G. L., Schlichtmann, U., & Li, B. (2024). LLM-aided efficient hardware design automation. *arXiv:2410.18582*.
2. Salah, K., & AbdelSalam, M. (2017). Smart auto-correction methodology using assertions and dynamic partial reconfiguration. In *2017 24th IEEE International Conference on Electronics, Circuits and Systems (ICECS)*. IEEE.
3. Salah, K. (2017). MCERTL: Mutation-based correction engine for register-transfer level designs. *International Journal of Electronics and Communication Engineering, 4*(11).
4. Salah, K. (2015). New trends in RTL verification: Bug localization, scan-chain-based methodology, GA-based test generation. In *Proc. of DVCON*.

5. Retrieved from https://dvcon-proceedings.org/wp-content/uploads/pre-silicon-debug-automation-using-transaction-tagging-and-data-mining.pdf
6. Islam, M. U., Sami, H., Gaillardon, P.-E., & Tenace, V. (2025). EDA-aware RTL generation with large language models. In *Proc. Design, Automation & Test in Europe Conf. (DATE), Valencia, Spain*. IEEE.
7. Hayes-Roth, F. (1985). Rule-based systems. *Communications of the ACM, 28*(9), 921–932.
8. Chandrashekar, G., & Sahin, F. (2014). A survey on feature selection methods. *Computers & Electrical Engineering, 40*(1), 16–28.
9. Bengio, Y., Courville, A., & Vincent, P. (2013). Representation learning: A review and new perspectives. *IEEE Transactions on Pattern Analysis and Machine Intelligence, 35*(8), 1798–1828.
10. Heidari, M., & Alizadeh, B. (2025). FixRTL: Auto-correction of multiple RTL bugs by a new feature burst clustering algorithm and mutation. *ACM Transactions on Design Automation of Electronic Systems, 30*(4), 56. https://doi.org/10.1145/3733238
11. Behnam, P., Malfeh, B. A., & Navabi, Z. (2014). Automatic correction of certain design errors using mutation technique. In *ETS 2014 (European Test Symposium), May 26–30*.
12. Behnam, P., & Alizadeh, B. (2015). In-circuit mutation-based automatic correction of certain design errors using SAT mechanisms. In *IEEE 24th Asian Test Symposium*.
13. De Millo, R., & Offutt, A. (1991). Constraint-based automatic test data generation. *IEEE Transactions on Computers, 17*(9), 900–910.
14. Retrieved from https://publications.waset.org/abstracts/80068.pdf
15. Serrestou, Y., Beroulle, V., & Robach, C. (2007). Functional verification of RTL designs driven by mutation testing metrics. In *Proceedings of the 10th Euromicro Conference on Digital System Design Architectures, Methods and Tools (DSD '07), Lubeck, Germany* (pp. 222–227). IEEE.
16. Sen, A., & Abadir, M. S. (2010). Coverage metrics for verification of concurrent SystemC designs using mutation testing. In *Proc. of the 15th IEEE International High Level Design Validation and Test Workshop, Anaheim, CA, USA* (pp. 75–81). IEEE.
17. Jia, Y., & Harman, M. (2011). An analysis and survey of the development of mutation testing. *IEEE Transactions on Software Engineering, 37*(5), 649–678.
18. Kurtz, B., Ammann, P., & Offutt, J. (2015). Static analysis of mutant subsumption. In *Software Testing, Verification and Validation Workshops (ICSTW), 2015 IEEE Eighth International Conference on* (pp. 1–10).
19. Hyunsook, D., & Rothermel, G. (2006). On the use of mutation faults in empirical assessments of test case prioritization techniques. *IEEE Transaction on Software Engineering, 32*(9), 733–752.
20. Just, R. (2014). The major mutation framework: Efficient and scalable mutation analysis for Java. In *Proceedings of the 2014 International Symposium on Software Testing and Analysis, ser. ISSTA 2014* (pp. 433–436). ACM.
21. Le, D., Alipour, M. A., Gopinath, R., & Groce, A. (2014). MuCheck: An extensible tool for mutation testing of Haskell programs. In *Proceedings of the 2014 International Symposium on Software Testing and Analysis, ser. ISSTA 2014* (pp. 429–432). ACM.
22. Mirshokraie, S., Mesbah, A., & Pattabiraman, K. (2015). Guided mutation testing for JavaScript web applications. *IEEE Transactions on Software Engineering, 41*(5), 429–444.
23. Oliveira, R., Alégroth, E., Gao, Z., & Memon, A. (2015). Definition and evaluation of mutation operators for gui-level mutation analysis. In *Software Testing, Verification and Validation Workshops (ICSTW), 2015 IEEE Eighth International Conference on* (pp. 1–10). IEEE.
24. Yao, X., Harman, M., & Jia, Y. (2014). A study of equivalent and stubborn mutation operators using human analysis of equivalence. In *Proceedings of the 36th International Conference on Software Engineering, ser. ICSE 2014* (pp. 919–930). ACM.

25. Papadakis, M., Malevris, N., & Kintis, M. (2010). Mutation testing strategies – A collateral approach. In *Proceedings of the 5th International Conference on Software and Data Technologies (ICSOFT)* (Vol. Vol. 2, pp. 325–328). SciTePress.
26. Behnam, P., & Alizadeh, B. (2015). In-circuit mutation-based automatic correction of certain design errors using SAT mechanisms. In *IEEE 24th Asian Test Symposium*. IEEE.
27. Cleve, H., & Zeller, A. (2005). Locating causes of program failures. In *Proceedings of the 27th International Conference on Software Engineering, St. Louis, Missouri, USA* (pp. 342–351). ACM.
28. Wong, W. E., & Qi, Y. (2009). BP neural network-based effective fault localization. *International Journal of Software Engineering and Knowledge Engineering, 19*(4), 573–597.
29. Wong, W. E., Wei, T., Qi, Y., & Zhao, L. (2008). A Crosstab-based statistical method for effective fault localization. In *Proceedings of the 1st International Conference on Software Testing, Verification and Validation (ICST), Lillehammer, Norway* (pp. 42–51). IEEE.
30. Mohamed, K. S. (2023). An introduction to deep learning. In *Deep learning-powered technologies: Autonomous driving, artificial intelligence of things (AIoT), augmented reality, 5G communications and beyond* (pp. 1–57). Springer Nature Switzerland.
31. Ouyang, L., Wu, J., Jiang, X., Almeida, D., Lowe, R., et al. (2022). Training language models to follow instructions with human feedback. *arXiv:2203.02155*.
32. Wang, A., & Cho, K. (2019). BERT has a mouth, and it must speak: BERT as a Markov random field language model. In *Proc. of the Workshop on Methods for Optimizing and Evaluating Neural Language Generation*. Association for Computational Linguistics.
33. Cao, Y., et al. (2023). A comprehensive survey of AI-generated content (AIGC): A history of generative AI from GAN to ChatGPT. *arXiv preprint arXiv:2303.04226*.
34. Huang, S., et al. (2023). Language is not all you need: Aligning perception with language models. *arXiv preprint arXiv:2302.14045*.
35. L. Tunstall, L. von Werra, and T. Wolf, Natural language processing with transformers: Building language applications with hugging face. O'Reilly Media, 2022.
36. Brown, T. B., Mann, B., Ryder, N., Subbiah, M., Kaplan, J., Dhariwal, P., Neelakantan, A., Shyam, P., Sastry, G., Askell, A., Agarwal, S., Herbert-Voss, A., Krueger, G., Henighan, T., Child, R., Ramesh, A., Ziegler, D. M., Wu, J., Winter, C., Hesse, C., Chen, M., Sigler, E., Litwin, M., Gray, S., Chess, B., Clark, J., Berner, C., McCandlish, S., Radford, A., Sutskever, I., & Amodei, D. (2020). Language models are few-shot learners. In *Proceedings of the 34th International Conference on Neural Information Processing Systems, ser. NIPS '20*. Curran Associates.
37. Rozière, B., Gehring, J., Gloeckle, F., Sootla, S., Gat, I., Tan, X. E., Adi, Y., Liu, J., Sauvestre, R., Remez, T., Rapin, J., Kozhevnikov, A., Evtimov, I., Bitton, J., Bhatt, M., Ferrer, C. C., Grattafiori, A., Xiong, W., Défossez, A., Copet, J., Azhar, F., Touvron, H., Martin, L., Usunier, N., Scialom, T., & Synnaeve, G. Code Llama: Open foundation models for code. *arXiv:2308.12950v3 2024*.
38. Nakkab, S., Zhang, Q., Karri, R., & Garg, S. (2024). Rome was not built in a single step: Hierarchical prompting for LLM-based chip design. In *Proceedings of the 2024 ACM/IEEE International Symposium on Machine Learning for CAD, ser. MLCAD '24*.
39. Lu, Y., Liu, S., Zhang, Q., & Xie, Z. (2023). RTLLM: An open-source benchmark for design RTL generation with large language model. *arXiv:2308.05345*. https://arxiv.org/abs/2308.05345
40. Chen, X., Lin, M., Schärli, N., & Zhou, D. (2023). Teaching large language models to self-debug. *arXiv preprint arXiv:2304.05128*.
41. Fu, Y., Zhang, Y., Yu, Z., Li, S., Ye, Z., Li, C., Wan, C., & Lin, Y. C. (2023). GPT4AIGChip: Towards next-generation ai accelerator design automation via large language models. In *IEEE/ACM International Conference on Computer Aided Design (ICCAD)* (pp. 1–9). IEEE.
42. Baartmans, R., Ensinger, A., Agostinelli, V., & Chen, L. (2025). ML for hardware design interpretability: Challenges and opportunities. *arXiv preprint arXiv:2504.08852*. https://arxiv.org/abs/2504.08852

43. Zhao, Y., Zhang, H., Huang, H., Yu, Z., & Zhao, J. (2024). MAGE: A multi-agent engine for automated RTL code generation. https://doi.org/10.48550/arXiv.2412.07822
44. Xu, K., Sun, J., Hu, Y., Fang, X., Shan, W., Wang, X., & Jiang, Z. (2024). MEIC: Re-thinking RTL debug automation using LLMs. In *Proceedings of the 43rd IEEE/ACM International Conference on Computer-Aided Design* (pp. 1–9). ACM.
45. Wang, N., Yao, B., Zhou, J., Hu, Y., Wang, X., Guan, N., & Jiang, Z. (2025). VeriDebug: A unified LLM for Verilog debugging via contrastive embedding and guided correction. https://doi.org/10.48550/arXiv.2504.19099
46. Thakur, S., Ahmad, B., Fan, Z., Pearce, H., Tan, B., Karri, R., et al. (2023). Benchmarking large language models for automated verilog RTL code generation. In *2023 Design, Automation & Test in Europe Conference & Exhibition (DATE)* (pp. 1–6). IEEE.
47. Liu, M., Pinckney, N., Khailany, B., & Ren, H. (2023). Verilogeval: Evaluating large language models for verilog code generation. In *2023 IEEE/ACM International Conference on Computer Aided Design (ICCAD)* (pp. 1–8). IEEE.
48. Huang, H., Lin, Z., Wang, Z., Chen, X., Ding, K., & Zhao, J. (2024). Towards LLM-Powered Verilog RTL Assistant: Self-Verification and Self-Correction. *arXiv:2406.00115*. https://arxiv.org/abs/2406.00115
49. ul Islam, M., Sami, H., Gaillardon, P.-E., & Tenace, V. (2024). AIvril: AI-driven RTL generation with verification in-the-loop. *arXiv:2409.11411v1*. https://arxiv.org/abs/2409.11411
50. Huang, H., Lin, Z., Wang, Z., Chen, X., Ding, K., & Zhao, J. (2024). Towards LLM-powered Verilog RTL assistant: Self-verification and self-correction. *arXiv preprint arXiv:2406.00115.*
51. Qiu, R., Zhang, G. L., Drechsler, R., Schlichtmann, U., & Li, B. (2024). CorrectBench: Automatic testbench generation with functional self-correction using LLMs for HDL design. *arXiv preprint arXiv:2411.08510.*
52. Qiu, R., Zhang, G. L., Drechsler, R., Schlichtmann, U., & Li, B. (2025). VeriReason: Reinforcement learning with testbench feedback for reasoning-enhanced verilog generation. *arXiv preprint arXiv:2505.11849.*
53. Ahmad, B., Ah-Kiow, J., Tan, B., Karri, R., & Pearce, H. (2025). FLAG: Finding line anomalies (in RTL code) with generative AI. *ACM Transactions on Design Automation of Electronic Systems (TODAES), 30*(5), Article 56. https://doi.org/10.1145/3736411
54. Ghosh, R., Lin, M., Li, X., Xu, Z., Kumar, R., & Malik, S. (2025). Spec2Assertion: Generating SystemVerilog assertions from natural language specifications using large language models. *arXiv preprint arXiv:2505.07995*. https://arxiv.org/abs/2505.07995
55. Tsai, Y. D., Liu, M., & Ren, H. (2023). RTLFixer: Automatically fixing RTL syntax errors with large language models. *arXiv preprint arXiv:2311.16543.*
56. Yao, X., Li, H., Chan, T. H., Xiao, W., Yuan, M., Huang, Y., Chen, v L., & Yu, B. (2024). HDLdebugger: Streamlining HDL debugging with Large Language Models. *arXiv preprint arXiv:2403.11671.*
57. Weimin, F., Yang, K., Dutta, R. G., Guo, X., & Qu, G. (2023). LLM4SecHW: Leveraging domain-specific large language model for hardware debugging. In *2023 Asian Hardware Oriented Security and Trust Symposium (AsianHOST)* (pp. 1–6). IEEE.
58. Fang, W., Li, M., Li, M., Yan, Z., Liu, S., Zhang, H., & Xie, Z. (2024). AssertLLM: Generating and evaluating hardware verification assertions from design specifications via multi-LLMs. *arXiv preprint arXiv:2402.00386.*

Index

K. Mohammed, *The Art of RTL Debugging*, Synthesis Lectures on Digital Circuits & Systems, https://doi.org/10.1007/978-3-032-15880-2

MIX
Papier aus verantwortungsvollen Quellen
Paper from responsible sources
FSC® C105338

If you have any concerns about our products,
you can contact us on
ProductSafety@springernature.com

In case Publisher is established outside the EU,
the EU authorized representative is:
Springer Nature Customer Service Center GmbH
Europaplatz 3, 69115 Heidelberg, Germany

Printed by Libri Plureos GmbH
in Hamburg, Germany